AIAA Aerospace
Assessment Series
Volume 2

Proceedings of the Symposium
University of Santa Clara
Santa Clara, California
June 19-23, 1978

Utilization of Alternative Fuels for Transportation

Sponsored by
United States Department of Energy

Organized and Operated by
American Institute
of Aeronautics and Astronautics
and University of Santa Clara

Edited by
Martin Newman and Jerry Grey
April 15, 1979

American Institute
of Aeronautics and Astronautics
1290 Avenue of the Americas
New York, New York 10019

American Institute of Aeronautics and Astronautics
New York, New York

Library of Congress Cataloging in Publication Data
Main entry under title:

Utilization of alternative fuels for transportation.

(AIAA aerospace assessment series; v.2)
Bibliography: p.
1. Fuel—Congresses. I. Newman, Martin. II. Grey, Jerry. III.
United States Dept. of Energy. IV. American Institute of Aeronautics
and Astronautics. V. Santa Clara, Calif. University. VI. Series:
American Institute of Aeronautics and Astronautics. AIAA aerospace
assessment series; v. 2.

TP315. U847	629.04	79-12149

ISBN 0-915928-31-0

AIAA Aerospace
Assessment Series
Volume 2

Proceedings of the Symposium
University of Santa Clara
Santa Clara, California
June 19-23, 1978

Utilization of
Alternative Fuels
for Transportation

Sponsored by
United States Department of Energy

Organized and Operated by
American Institute
of Aeronautics and Astronautics
and University of Santa Clara

Edited by
Martin Newman and Jerry Grey
April 15, 1979

American Institute
of Aeronautics and Astronautics
1290 Avenue of the Americas
New York, New York 10019

American Institute of Aeronautics and Astronautics
New York, New York

Library of Congress Cataloging in Publication Data
Main entry under title:

Utilization of alternative fuels for transportation.

(AIAA aerospace assessment series; v.2)
Bibliography: p.
1. Fuel—Congresses. I. Newman, Martin. II. Grey, Jerry. III.
United States Dept. of Energy. IV. American Institute of Aeronautics
and Astronautics. V. Santa Clara, Calif. University. VI. Series:
American Institute of Aeronautics and Astronautics. AIAA aerospace
assessment series; v. 2.
TP315. U847   629.04   79-12149
ISBN 0-915928-31-0

FOREWORD

"It will be recorded in history that of all the world's nations, ours used most of, and benefited most from, the riches of petroleum that were drained from the Earth. Will it also be recorded that we used part of our resulting industrial might, as the world's petroleum supplies diminished, to provide an attractive alternative?"

Richard Pefley
(from opening statement,
Santa Clara, Calif.,
June 20, 1978)

This Symposium is our attempt at an affirmative answer.

TABLE OF CONTENTS

*See Section II and Appendix A for authors' affiliations.

PREFACE

The purpose of this Symposium was to assemble the principal practitioners and potential users of alternative fuel technology in order to exchange information and review and assess the status of and prospects for the utilization of alternative fuels in transportation.

The Symposium was structured to include a large measure of informality. Among a gathering of peers, this was considered the most effective way of getting at any ideas provoked. The tone was set from the start, with questions and observations from the panel and the floor following the keynote; and the evening workshops, by design, were by and large free-for-all exchanges.

Although the session sequence as it actually occurred at the Symposium has been reordered in these Proceedings to provide the necessary subject continuity, the contents of each session are presented here in something close to the order in which they took place. Unless otherwise noted, all summaries of participants' prepared remarks (as opposed to formal papers) are exactly that—summaries. In no respect should they be construed as quoted or paraphrased. They are attempts by the Editors to express essential meanings as gleaned from audio tapes of the Proceedings.

Other than in the Summary sections, the opinions expressed in this volume do not necessarily constitute a consensus of the gathering; together they are simply a compendium of individual points of view.

Martin Newman
Jerry Grey

January 8, 1979

I.
SUMMARY OF CONCLUSIONS

A. Energy Supply and Fuel Manufacturing/Processing

Our present domestic supply of petroleum is inadequate to meet national requirements. Our need to import enough petroleum to fill the gap is creating a dangerous balance-of-payments problem. Beyond that, the time approaches when even the world supply will be inadequate. The job of the technologist, then, is to develop workable petroleum substitutes. The short-term goal is to replace imported petroleum; the long-term goal is to replace petroleum itself. Of the possible new materials we have to work with, the likeliest are coal, oil shale, and agricultural resources (biomass).

At this stage in our technology for converting these materials, we need empirical field experience in order to (a) test the technology; (b) establish costs; (c) establish acceptability of products; and (d) establish and demonstrate methods for attaining environmental acceptability of products. And here the public becomes involved.

The public's involvement in alternate-fuel development may properly be a subject for the "Impacts and Institutional Issues" section of this summary. But to a large degree it affects the directions of the technological work and should therefore be touched upon here. There are important differences between energy companies and segments of the public in the matters of priorities and strategies. The major energy companies, for their part, are concentrating on getting the maximum use of existing gas and petroleum deposits while they devote their alternate-fuel efforts toward converting coal and shale to gaseous and liquid fuels, and on doing this as economically as possible, i.e., producing the largest quantities of fuels for the least cost. The major difficulties with this strategy relate to the public. To begin with, as to public interest in the work, there is little: the nature of the activities is highly technical. As to visibility, it is (a) limited, because the activities are concentrated, and (b) negative, because there is high local and environmental impact.

The public priority, to the extent that there is one, comes from particular quarters: agricultural interests and concerned public-interest groups. Their choice for an alternative fuel is the

1

agriculturally based alcohols: ethanol from agricultural products and wastes (to be utilized as in Nebraska's "Gasohol" program), and methanol, via gasification of forest and agricultural wastes. High among these groups' strategies and priorities are (a) the short-term use of agricultural surplus to produce grain alcohol; (b) its initial use in gasoline blends; and (c) the meeting of entire costs with either substantial state subsidy (as in the case of California) or a campaign for public acceptance at a higher price (as in Nebraska). The major difficulties here, in addition to the higher costs, are the geographical limitations on the use of alcohol, at least to begin with, and its limited contribution to the overall energy problem in the near term and the medium term.

As for other alternatives, hydrogen and electricity are technologically feasible. Hydrogen, in particular, is of interest as a fuel for long-range aircraft. In the longer term, when conventional jet fuel would be made synthetically, hydrogen could be a competitive alternative. Also in the longer term, should CO_2 emissions become enough of a problem, electrolytic hydrogen could be a viable though expensive option. Electricity, in battery-powered ground vehicles, is expected to have growing use; but when compared for the near and medium terms with hydrocarbon-fueled or methanol-powered heat engines, its performance is limited and its costs are high.

Our primary resources for alternative hydrocarbon fuels are coal and oil shale. Synthetic fuel derived from shale can be expected to match easily the characteristics of today's fuels. However, our major resource by far is coal, and coal tends to produce fuels of low hydrogen content. These would be high-octane fuels, and their use in spark-ignition engines would result in improved fuel efficiency and manufacturing economy. However, the picture is less attractive if we look to use them in conventional jet and diesel engines—which require fuels of high hydrogen content. In such cases, hydrogen would have to be added, which would add substantially to the cost and to the energy consumed during manufacture. A logical attack on the problem is from the other end. For example, extensive work is being done to develop jet engines that can accept low-hydrogen-content fuel.

B.   Storage and Distribution

1.   Cryogenic Fuels

(a)   Hardware can be built with today's technology that will satisfy most ground requirements for large-scale storage and distribution of both liquid hydrogen and liquid methane. Technology advances are required only to improve efficiency (and lower cost).

(b) Capital requirements to supply liquid hydrogen or liquid methane to any major transportation mode would be high.

(c) The fuel implementation schedule would be dictated by industry's capacity to build and install storage distribution equipment.

(d) The safety of large-scale cryogen storage and distribution systems must be reviewed.

2. Synthetic Fuels

(a) Synthetic hydrocarbons that resemble today's gasoline components would present few problems to the existing transportation and distribution network.

(b) The low-molecular-weight ethers, as potential gasoline-blending components, likewise present few problems.

(c) Methanol and, to a lesser extent, ethanol blended with gasoline would present serious problems of phase separation due to water contamination in our existing "wet" distribution system. This could be avoided by transporting and distributing them in their pure form and blending them at the point of sale, but would require segregated facilities, logistical costs, and blending facilities. The total additional costs would have to be weighed.

C. Utilization

(a) Changes in fuel and engine utilization will be evolutionary rather than revolutionary. There is too much "inertia" built into the present fuel and engine systems for it to be otherwise. This does not preclude the possibility of a "revolution" in a given subsystem. For example, if liquid hydrogen were to be used as an aircraft fuel, the entire plane and the fuel system would have to be changed.

(b) Present engines will continue to dominate the field: spark-ignition engines for automobiles and diesel engines for trucks. However, there will probably be some penetration of the gas turbine into the heavy-duty-truck field; some further penetration of the diesel into the passenger-car field; and increased use of turbocharging.

(c) To increase the use of alternative fuels will require primarily engineering changes rather than scientific breakthroughs. Techniques for burning alcohol in spark-ignition engines are known. Alternate hydrocarbon fuels can be made compatible with existing engines. Engine developments might permit use of somewhat different hydrocarbon fuels in order better to utilize the fuel raw material. Direct use of powdered, refined coal in reciprocating engines would require development of new engines.

(d)   Aside from societal considerations, it is necessary, because of the inertia and delays in the system, to start *now*. It will take at least 10 to 15 years just to begin commercial production, and several times that long to switch completely.

D.   Impacts and Institutional Issues

(a)   We are witnessing small beginnings everywhere, from action in Congress to highly localized events—small state-by-state, city-by-city attempts to pull together resources to meet some kind of a fundamental need in an arena where there is high public interest and very high public and private risk. In the best sense these are real social experiments—they wed technology with public need, and they involve as actors politicians, bureaucrats, entrepreneurs, technologists, financial people, and consumers.

(b)   The technological hurdles in getting to an alternative fuel system are falling much more quickly than the institutional ones. These must be similarly analyzed, attacked, and felled if we expect to shorten the long time constants for change in so complex and unwieldy a system as energy. We are only just beginning to understand what the institutional hurdles are [see (f) below]. This is an undertaking requiring long-range government planning and conscious action by major institutions.

(c)   The biomass experiments are important symbols. Taken together with certain state-by-state, city-by-city, entrepreneur-by-entrepreneur activities, they signify that things are beginning. This is important because the government cannot begin things by itself. In order to be an effective instrument of change, it must represent large constituencies in many places. The biomass movement, simply because it is happening, is very encouraging. As for the forces driving the movement, these must be examined and considered. There are messages here that are not yet clear, but they are here.

(d)   We must have well-designed demonstration projects of the sizes that we expect will become commercial facilities for all the technologies. Just as we test and prove out a technological apparatus, so do we have to prove out each institutional apparatus. We must, for example, look at the flow of money in and out of the facility, the capital formation, how the facility gets constructed, how operated, how it fails, what its maintenance difficulties are. And to get useful answers, the prototypes must be entire plants, each serving as a nucleus for an industry.

(e)   We must establish some kind of aggressive, nonconventional liaison across all of the institutions that have to have a stake in the outcome of this effort.

(f)   The following is a listing of some of the issues that must be responded to in the national policy, and must be dealt with, one way or another, in our moving to very large new industries and complex facilities of exotic technologies:

(i)   Financing and capital availability.

(ii)   Facility siting (and the permitting systems for siting).

(iii) Air quality, all along the fuel cycle—at conversion sites as well as end-use locations. This consideration clearly interacts with facility siting.

(iv)   Water availability for cooling and as a source of hydrogen.

(v)   Reclamation of strip-mined lands and coping with subsidence from underground mining.

(vi)   Rapid urban growth (boom towns) in areas previously not industrialized.

(vii) Optimizing energy flows from competing technologies—i.e., capturing the greatest amount of product energy per dollar input as well as per unit-of-resource input; and comparing these inputs for various end uses to determine the most efficient energy investment as well as economic investment.

(viii) Resource depletion, i.e., coal and oil shale. Conservative estimates based upon careful studies of projected growth patterns, plus the projected establishment of large coal-based industries, indicate that within 75 years, easily 50% of the total recoverable coal reserves would be either mined or encumbered on long-term contracts; and that the rate of use at that point would be so huge that 80% of the available U. S. cheap coal would be gone before 2100.

(ix)   Inter-regional shifts in coal production and the growth patterns resulting therefrom. To a great degree, these shifts will be influenced by whatever national energy strategies come out of Congress, how they are pushed by the Department of Energy, and how vigorously their environmental controls are enforced by the Environmental Protection Agency.

(x) Effects on ecosystems and human population. For example, a great deal of research is still needed to learn the effects of and solve any problems caused by the accumulations in the atmosphere of carbon dioxide from combustion; emissions of SO_2 in the midwest and the chemical changes it undergoes as it moves eastward to descend in rainfall on the northeast; emissions resulting from vastly increased coal use in regions formerly dependent on natural gas or fuel oil.

(xi) Disruptions in the marketplace during the transition period from traditional fuels to coal and coal products.

(xii) The need to adjust demand in order to produce a measure of symmetry with supply, i.e., how to soften the environmental impacts on the supply side by lessening the demand for synthetic fuel.

(g) The foundation on which all of the above must be built is public credibility: a public "gut" feeling that the institutions involved—private and public—are saying what they really mean, and will do what they say. Without that there will be no real support for the effort, and the effort will fail. The vital importance that the effort succeed requires that we overcome the current low esteem of industry and government with the public. Active and visible involvement of credible, broad-based, public-interest institutions will be necessary from the start. And the start cannot be delayed. What we do or do not do now sets the pace for what happens in alternative fuels during the next 20 years.

II.
SYMPOSIUM ORGANIZATION

The initial steps in conceiving and initiating the organization of the Symposium were taken by E. Eugene Ecklund of the U.S. Department of Energy. He convened the Steering Committee named below to formulate and implement the program. The American Institute of Aeronautics and Astronautics undertook to administer the Symposium under DOE sponsorship, and the University of Santa Clara volunteered to serve as the Symposium's host.

These Proceedings, whose raw material was generated at the Symposium, have been edited and published by the AIAA. All the Symposium participants listed in Appendix A, however, were given an opportunity to comment on and suggest revisions to the text prior to publication.

SYMPOSIUM STEERING COMMITTEE

Richard Alpaugh
Non-Highway Systems Branch,
Div. of Transportation Energy
Conservation,
U.S. Department of Energy

John M. Bailey
Caterpillar Tractor Company

E. Eugene Ecklund
Alternative Fuels Utilization Branch,
Div. of Transportation Energy
Conservation,
U.S. Department of Energy

William Escher
Escher Technology Associates

John F. Freeman
Sun Company

Conan P. Furber
Association of American Railroads

Jerry Grey
American Institute of Aeronautics
and Astronautics

Graham Hagey
Div. of Technology Overview,
U.S. Department of Energy

Jerome Hinkle
Div. of Technology Overview,
U.S. Department of Energy

F.H. Kant
Exxon Research and Engineering
Company

Dan Maxfield
Div. of Transportation Energy
Conservation,
U.S. Department of Energy

A.M. Momenthy
Boeing Commercial Airplane
Company

Andrew Parker
Mueller Associates, Inc.

Richard Pefley
University of Santa Clara

Roy D. Quillian
Southwest Research Institute

Earl Van Landingham
Office of Energy Programs, NASA

SESSION AND WORKSHOP CHAIRMEN

Energy Supply and Fuel Manufacturing/Processing
John P. Longwell
Massachusetts Institute of Technology

Storage and Distribution
Jack H. Freeman
Sun Oil Company
Albert M. Momenthy
Boeing Commercial Airplane Company

Road Vehicle Utilization and Nonroad Vehicle Utilization
Philip S. Myers
University of Wisconsin

Impacts and Institutional Issues
Jerome Hinkle
U.S. Department of Energy

Summary
E. Eugene Ecklund
U.S. Department of Energy

CONFERENCE ADMINISTRATION

Jerry Grey
AIAA (Administrative Director)
Walter Brunke
AIAA

Richard Pefley
University of Santa Clara
Tom Timbario
Mueller Associates, Inc.

III.
ENERGY SUPPLY AND FUEL MANUFACTURING/PROCESSING

A. Future Transportation Fuels

Martin R. Adams
U. S. Department of Energy
Washington, D. C.

1. Abstract

This paper presents an overview of the expected future need for alternative transportation fuels. It includes an assessment of the potential of conservation technologies for reducing future fuel demand. The institutional barriers that may impede the introduction of new transportation fuel systems into the national infrastructure are described and a discussion is provided of the expected trend of shifting from primarily gasoline consumption to a preponderance of consumption of middle distillates during the remainder of this century. An overview of the technological, economic, and environmental/safety status of various alternative fuel candidates is presented, and the paper closes with a discussion of government roles in accelerating the adoption of new technologies.

2. Introduction

Figure 1 displays the 1975 U. S. energy consumption according to consuming sector. The transportation sector's share of the total has remained relatively constant over the past 25 years at approximately one fourth of the total as shown in Figure 2. For nearly the last two decades, over 95% of total transportation energy has been provided by petroleum products, and transportation needs have consumed over 50% of all petroleum imported and produced domestically. The finite crude oil resource base, currently the source of nearly all transportation fuel, is being rapidly depleted. This clearly indicates the requirement for alternative fuels for transportation. Further, given the expected 20-40 years required to deploy new technologies, the nation may be beginning to fall behind the "power curve" in developing new supply options.

In 1976 the transportation sector used nearly 19 Quads or 26% of the nation's energy consumption. Nearly all of the fuel consumed was liquids and most of the fuel was consumed by highway vehicles.

9

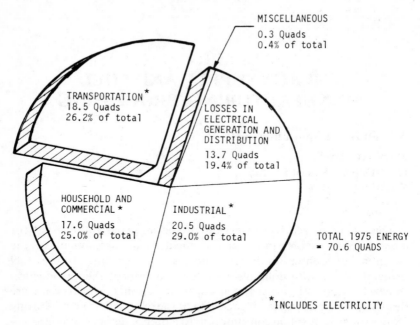

MISCELLANEOUS
0.3 Quads
0.4% of total

TRANSPORTATION*
18.5 Quads
26.2% of total

LOSSES IN
ELECTRICAL
GENERATION AND
DISTRIBUTION

13.7 Quads
19.4% of total

HOUSEHOLD AND
COMMERCIAL *

17.6 Quads
25.0% of total

INDUSTRIAL *

20.5 Quads
29.0% of total

TOTAL 1975 ENERGY
= 70.6 QUADS

*INCLUDES ELECTRICITY

Fig. 1   1975 U.S. energy consumption by sector[1]

ERDA studies have projected that this total consumption will grow to approximately 38 Quads/year in 2010, even under the assumption that the automobile industry will successfully meet the mandated fleet average fuel economy requirement of 27.5 miles per gallon for new cars beginning with the 1985 models. The successful market penetration of new transportation technologies, especially advanced heat propulsion engines, over the 35 year period was determined by these ERDA studies to reduce this 2010 projected level of consumption to about 25 Quads/year. This is still some 30% above today's level. It is significant that, in some of these penetration studies, total fuel consumption in the transportation sector remained virtually constant between 1975 and 2000 due to the penetration of the new, more efficient heat engines and other improved technologies. Fuel consumption began to increase rapidly after 2000 due to saturation of the benefits of the most efficient technologies.

As remaining domestic oil production becomes progressively dependent on enhanced oil recovery and production from frontier areas (Alaska, deep onshore; Lower-48, offshore), and with much overseas production subject to arbitrary OPEC price increases, oil prices can be expected to continue increasing in real terms. Consequently, alternative technologies, even without government sup-

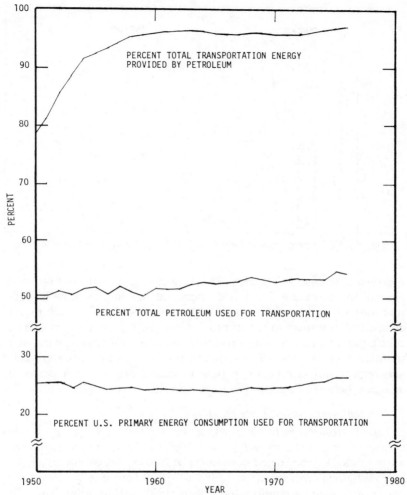

Fig. 2   1950-1976 energy use by transportation sector[2]

ports or controls, may become economically competitive in the future. This portends, however, higher future energy prices in real terms. From the point of view of the individual, the more one must pay for personal transportation, the less one has available for alternative uses of personal funds, without offsetting increases in real income. American lifestyles can be characterized by high elective mobility decisions, that is, selection of locations for domiciles with only minimal regard to commuting distances to places of employment, ability to travel extensively for business or personal reasons at low cost, and so forth. A major domestic industry has developed

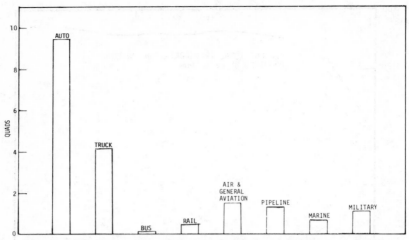

Fig. 3   1975 energy consumption in Quads by transportation mode[4]

specifically to cater to travelers. If future costs of transportation fuels absorb an increasing fraction of personal income, eventually consumption of transportation fuels and services, which has been rather inelastic with respect to recent price increases, will be affected. Thus, substantial real increases in transportation costs will have obvious and significant impacts on the national economy and on lifestyles. This may require shifts in the way people allocate their disposable personal incomes, between housing, food, transportation, recreation, etc.

3. Potential Impact of Conservation

The distribution by mode of energy used in the U. S. transportation system in 1975 is shown in Figure 3. Highway users, in particular the automobile, are the largest consumers of transportation energy. Total highway users accounted for 73% of 1975 consumption, and the automobile alone accounted for 51% of 1975 consumption. Thus the automobile has been a particular target of efforts to reduce transportation energy demand, and much private and government-assisted R&D has been oriented toward improving its fuel economy.

Besides the automobile, trucks, aircraft, and pipelines are the other large consumers of transportation energy. However, it seems reasonable for the government to direct relatively less effort toward these end users, for the following reasons:

(a) All are relatively small energy users compared to the automobile.

(b) Small trucks principally used in urban areas will benefit indirectly from work on electric vehicles and on advanced engines for cars.

Table 1 Energy Intensity of Urban and Interurban Passenger Transportation Modes [3]

	Urban (Btu/passenger mile)	Interurban (Btu/passenger mile)
Cars	6600	3400
Buses	400-1000 [a]	1600-2900
Heavy Rail	230-900 [a]	450
Airplanes	Not Applicable	8400

[a] Lower number corresponds to peak periods, higher number to off-peak periods.

(c) Pipelines, interurban trucks, and airlines are operated by profit-oriented businesses with relatively strong incentives to minimize their consumption of energy. There is ample evidence that these industries are responding prudently to rising energy prices.

Automobiles also consume more energy per passenger mile than other transportation modes (except aircraft) as shown in Table 1. While Table 1 suggests it would be desirable to encourage the shifting of passengers to more efficient modes of transportation, such shifts in the near term are unlikely on a scale that would save significant amounts of energy. Thus, a more efficient automobile is probably the key to a more efficient transportation system.

This is not to say, however, that improvements in other areas are unimportant. ERDA's MOPPS* study which was conducted in 1977, examined the impact of the market penetration of new transportation techniques and technologies over the period 1975 to approximately 2010. Transportation technologies which were considered are listed and described in Table 2. Though not shown on the table, electrically powered vehicles provide an alternative to utilization of conventional transportation fuels, and are a target for developing and demonstrating improved technology. Estimates of the demand for transportation energy to the year 2010 and the savings that might be achieved through implementation of conservation technologies are shown in Figure 4.

Savings resulting from the market penetrations of conservation technologies were projected to be over and above the anticipated savings from the mandated 1985 average mileage requirement. Annual demand is projected to grow to approximately 38 Quads in 2010 without conservation technologies and approximately 25 Quads with their successful implementation. Figure 4 shows that the major conservation savings derive from improvements in heat engine propulsion systems.

---

*The Market Oriented Program Planning Study (MOPPS) conducted by the U. S. Energy Research & Development Administration (ERDA) in 1977 assumed the policy framework of the proposed National Energy Act (NEA).

Table 2   Generic Energy Conservation Technologies[4]

Mode	Generic Technology	Description
Highway (Auto, Bus, and Truck)	Waste Heat Utilization	Bottoming cycle application for trucks and regenerative braking.
	Chassis/Body	Improved design to reduce aerodynamic drag and weight reductions.
	Heat Engine Propulsion	Improved accessory drives, continuously variable transmissions, and new heat engines.
	Operating Techniques	Driver conservation training, diesel engine rerating, car pools, van pools, and vehicle component improvements.
Rail	Propulsion Systems	Idle cutoff, bottoming cycle, alternative switching, and electrification.
	Systems and Components	Lightweight cars, wheel slip systems, and streamlining.
	Operating Procedures	Reduction in empty backhauls, improved yard operation, and elimination of caboose.
Aviation	Engine and Airframe Design	Consideration of long-, medium-, and short-range aircraft as replacement aircraft in subsequent years as predicted by the NASA Aircraft Fuel Conservation Technology Program.
	System Design and Operation	Improved operating procedures and the training of pilots in fuel-saving procedures in the general aviation sector, including parabolic flight path trajectories.
Pipeline	Pumping Technologies	Friction reduction techniques, turbo-compounding reciprocating engines, and bottoming cycles.
	System Design and Operation	Control system optimization.
Marine	Powerplants and Auxiliary Systems	Adiabatic diesel, high-pressure reheat steam, low-speed diesel, and nuclear propulsion.
	System Design and Operational Improvements	More frequent hull cleaning, reduction of auxiliary loads, reduction of steam leaks, and operating at design conditions.
	Propulsors	Improved means of maneuvering or propelling ships.
	Hydrodynamics	Improved hull shapes and surface effect technologies.
	Structural Improvements	Use of high-strength steel and aluminum to lighten vessels.

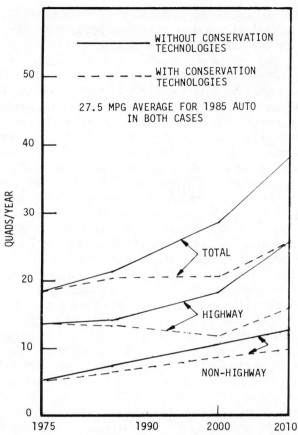

Fig. 4   Transportation energy demand with and without implementation of conservation technologies[4]

It should be noted that the major automobile manufacturers, after reviewing these projections, considered them to be optimistic for various reasons. They expressed considerably less optimism than this study about the impact of conservation technologies and were dubious of improvements in fleet average mileage beyond 27.5 mpg in the absence of large fuel price increases up to about $7-8/million Btu.

4.   Institutional Barriers to Commercialization of Alternative Fuels

A massive transportation system consisting of highways, rail lines, waterways, pipelines, and air transportation systems has evolved in the United States to meet the nation's local, regional, and national transportation needs. Correspondingly, an extensive infrastructure of import facilities, refineries, distribution networks, and retail outlets has developed to provide fuels for the transportation sector. While this sytem has evolved to a state of high operating efficiency, its size and complexity has the disadvantage of making it rather inflexible. Additionally, the automobile industry and other industries which supply vehicles for the transportation sector have large-scale in-place investments in manufacturing facilities built to produce vehicles with engines designed to burn particular fuels. Thus, the capability to produce an alternative liquid fuel at a price comparable, at some point in time, to that of gasoline, is a necessary but not sufficient condition for commercialization. Logistical problems in distribution and retailing must be solved, which may require further economic incentives. Such problems are magnified by the number of alternative fuels which could enter the market during the same time frame.

Thus, while retail outlets were able to recently incorporate facilities for unleaded gasoline relatively smoothly, these outlets might become hard pressed to add facilities for automotive diesel fuel, alcohol or alcohol/gasoline blends, and charging equipment for electric vehicle batteries, in addition to their facilities for conventional gasoline. If incompatibilities with vehicle engines exist which can be rectified through engine modifications, then procedures and incentives must be established for the manufacturers to incorporate the modifications for some or all of their production. Transitional problems may be severe, and introduction of new fuels and distribution networks may be resisted by existing suppliers. A recent example of this is the railroads' opposition to coal slurry pipelines. Thus, while demonstration of technological and economic feasibility for producing an alternative fuel is the major milestone, it is only a step along the path toward commercial implementation. The remaining steps will require the resolution of institutional, environmental, sociopolitical, and legal issues and can be expected to involve environmental impact statements, litigations, and various delays which tend to increase costs of particular projects. The government can and must assist in the commercialization of alternate fuels through enactment of regulatory or incentive legislation, but often the passage of such legislation is difficult and time consuming.

5.   The Transition from Gasoline to Middle Distillate Consumption

Gasoline is by far the transportation fuel most heavily demanded in the U. S. today. Roughly six times as much gasoline (in Quads) was consumed in 1975 as was either of the next most heavily demanded fuels, diesel and jet fuel. However, various studies project a major shift for the remainder of this century toward middle distillates (e. g., diesel fuel, aviation fuel, kerosene) and away from motor gasoline. Demand for gasoline for automobiles is projected to decline after 1979

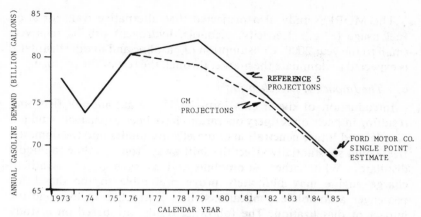

Fig. 5   Comparison of annual gasoline demand projections for passenger cars to 1985[5]

(Figure 5), reflecting the dominant effect of improvements in fuel economy through the 1985 models. This decline may taper off after 1985 and demand thereafter begin to increase again (not shown on the figure) due to the *net* effect of the following factors:

(a) Cars sold between 1990 and 1995 and having 1985 fuel economies would predominately replace cars of 1975-1980 vintage which are less efficient. This would tend to reduce gasoline demand.

(b) Newer model cars are driven more miles per year than the ones replaced, consequently tending to increase gasoline demand.

(c) The total car population continues to grow since new car sales each year exceed retirement of older cars. This tends to increase gasoline demand as well.

While gasoline demand decreases, demand for middle distillates is expected to increase substantially through the rest of the century. This trend has been projected by several sources (e. g., Refs. 6 and 7), although the extent of the shift differs between studies. MOPPS[6] predicts that by the year 2000 demand for diesel fuel and for jet fuel will each exceed demand for gasoline. While the shift offers economic advantages to consumers, it may portend significant changes for intermediate suppliers, e. g., petroleum refiners. Current U. S. refineries have the flexibility to increase middle distillate outputs at the expense of gasoline, but there are limits to this flexibility and the ease with which such conversions can take place. Consequently, any assessment of the future markets for middle distillate fuels must be made relative to the capability of domestic refineries to satisfy total fuel demand economically, preferably without increasing imports of distillates and/or crudes and without significantly restricting the output of other refined products.

The MOPPS study also projected that alternative transportation fuel usage (e. g., electricity, alcohols, hydrogen) will be relatively small to the year 2000. Consumption of gasoline and middle distillates is expected to dominate the market through the rest of the century.

6. The Impact of Dieselization

Introduction of diesel cars into the U. S. automobile fleet and resulting impacts on refinery operations have been extensively studied. Accelerated future penetration of diesel automobiles into the domestic fleet could substantially affect the shift away from gasoline to middle distillates. While other automobile options such as the stratified charge engine may ultimately prove preferable to the diesel for passenger automobile applications, it is instructive to examine the impact of dieselization. The following results are based on a study recently completed under contract to DOE.[5]

The number of diesel automobiles in the U. S. currently represents an almost insignificant fraction of the total car population. A trend toward greater penetration of diesel cars in the U. S. started with the introduction of the diesel version of the imported Volkswagen Rabbit in 1976-1977 and has since continued with options from General Motors for 1977-1978. General Motors is currently offering diesel versions of the Oldsmobile Delta 88, 98, and Custom Cruiser models and will have a diesel Cadillac Seville available in early 1979. The rate of increase in diesel car sales in the U. S. in the last few years has been dramatic, with expected annual sales of close to 100,000 cars in 1978 from a level of around 10,000 cars in 1976. While it is uncertain whether this rate will continue,† the trend toward an increasing diesel car population in the U.S. is evident.

The current market for diesel cars is influenced by the diesel's higher fuel economy relative to comparable gasoline cars (average of 25% more miles/gallon for current generation diesels), lower fuel prices, and generally lower maintenance costs which offset a significantly higher first cost at least for the domestically produced cars.

The reasons behind the interest in diesel automobiles are diverse. The interest is, however, indicative of the change in the mix in automotive powerplant that can be expected in the future, i. e., from a predominance of gasoline driven cars in the current mix to one that includes engines using diesel and broadcut fuels. However, for the present the diesel automobile is the only alternative to its gasoline

---

†It will depend both on the future course of federal emission standards for cars (NO$_x$, particulates, polynuclear aromatics, etc.) as well as sustained consumer acceptance of diesels (dictated by the ease of availability of diesel fuel, the fuel price differential with respect to gasoline, maintenance expense, etc.).

counterpart at least in the large size cars. The impacts on refinery operations of several assumed penetration rates of diesel automobiles into the new car market were evaluated in the Ref. 5 study. The penetration scenarios were:

(a) A rate increasing linearly from 1976 and reaching 65% of new car sales by 1995 (i.e., at 3.25% per year). At this rate, the diesel car population would reach 45% of the total fleet mix in 1995.

(b) A rate increasing exponentially from 0.1% of new car sales in 1976 to 65% by 1995. In this case, the proportion of diesels would reach 16% of the total fleet in 1995.

(c) A very rapid penetration rate in which all new cars starting with the 1982 models were assumed to be diesel cars. With this assumed rate, diesel cars would constitute 41% of the total fleet mix as early as 1985. This is an extreme case, included to bound the analysis.

In each case, demand for the other middle distillates was assumed to increase in accordance with projections made in a Brookhaven National Laboratory study.[7] The critical factor that determines the capability of U. S. refineries to meet the demands for both gasoline and middle distillates in the three dieselization scenarios is the gasoline to the distillate yield ratio (G/D). According to various industry studies, the G/D ratio is a measure of the limit to which refineries can accommodate varying levels in the output of gasoline and middle distillates. In Figure 6 the trends in this ratio are plotted for each case along with the limits to which refineries can alter their processing capability without major operational changes, namely, 0.6 to 0.7 G/D ratio. Note that these figures represent total distillates which include nontransportation fuels such as heating oil.

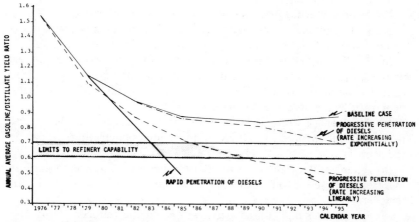

Fig. 6   Trends in gasoline/distillate yield ratios under different dieselization scenarios[5]

For the case of the high rate of diesel penetration (i. e., all new cars starting with the 1982 models being diesels), major changes in refining operations would be required by 1983-1984. This indicates that, should such a high rate of penetration of diesels occur, refineries must start planning significant operational changes as early as 1979-1980.

At an intermediate rate of diesel penetration (i. e., at a rate increasing linearly by 3.25% per year in 1976) the transition point occurs in 1985 at a G/D limit of 0.7, and in 1988 at a G/D limit of 0.6.

With a low initial rate of diesel penetration (i.e., at a rate increasing exponentially from 0.1% of new car sales in 1976 to 6.5% of new car sales in 1995), the transition point is reached in 1995 at a G/D limit of 0.7. When the G/D limit is lowered to 0.6, the transition point is extended beyond 1995 to approximately 2000.

A factor which could limit diesel penetration concerns the emission from diesel cars and the impact of future federal standards on these emissions. Current generation diesel automobiles have been shown capable of meeting both the 1977 Federal Standards and State of California Standards for hydrocarbons (HC), carbon monoxide (CO), and nitrogen oxides (NO_x). However, based on present technology, it does not appear feasible for diesel cars to meet a NO_x standard of 0.4 grams/mile without reducing their fuel economy or durability. In view of the superior fuel economy of diesel vehicles, it is possible that a variance might be given for the NO_x emissions, permitting levels between 1 and 1.5 grams/mile. In addition to NO_x, diesel cars emit a high level of particulates, a pollutant that is presently unregulated but likely to be covered in the future by standards. Modifications to diesel vehicles would be required to make them conform to these standards as well.

The study concluded that:

(a) No near-term (1985) refining problems are likely to result as a consequence of increasing diesel car population. This is because the penetration of diesel cars in this period is expected to be quite small.

(b) U. S. refineries with the flexibility to accomodate a modest increase in demand for diesel automotive fuel will have sufficient lead time to make significant changes to their operation, if the diesel penetrations are observed to increase rapidly.

(c) Stringent future federal emission standards for NO_x and particulates if enforced are likely to limit diesel penetration, at least in the near term.

(d) There will be savings in automotive fuel requirements with increasing dieselization, although these savings will be small in the near term.

(e) Barriers to high diesel penetration will be reduced if the fuel economy of new generation diesels is increased substantially.

7. Overview of Alternative Transportation Fuels

Several criteria have been suggested for assessing the merits of alternative transportation fuels. [8, 9] These criteria generally include the following assertions:

(a) The resource base for the energy source should be sufficiently extensive to supply synthesis plants over their lifetimes and provide fuel in sufficient quantities to meet demand.

(b) The fuel should be reasonably competitive with conventional fuels.

(c) The energy system including resource extraction, transportation, fuel production, distribution, and end-use should be operationally safe and environmentally acceptable in all phases.

(d) The energy system should have an overall efficiency as high as possible.

(e) The fuel should be reasonably compatible with existing and projected vehicular powerplants, have sufficiently high energy density to provide acceptable performance without an unacceptable size or weight penalty, and be convenient to use.

Figure 7 shows alternative paths in the transportation energy system from energy resource to transportation fuels. Primary energy sources (e.g., coal, sugar cane) must be transported to synthesis plants (e.g., coal liquefaction plants, fermentation and distillation plants) where they are converted to usable fuels, which are then distributed for intermediate assimilation and final sale.

A variety of candidate fuels can be produced from fossil resources and from biomass. These include: (a) distillate oils, (b) ethanol, (c) gasoline, (d) hydrogen, (e) methanol, (f) methane, and (g) propane.

Table 3 summarizes the chemical and physical properties of each of these fuels that are important considerations in assessing their value as transportation fuels. These transportation fuels may occur in liquid or

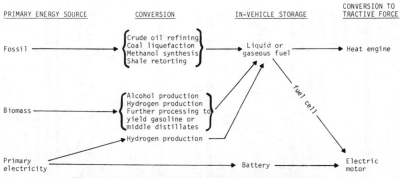

Fig. 7   Principal transportation energy options

Table 3 Fuel Properties[8]

Fuel	Chemical Formula	Lower Heating Value (Btu/lb)	Ignition Temperature (°F)	Tankage Weight (lb)[a]	Tankage Volume (gal)[a]	Engine Compatibility (Conventional Engine)	Large-Scale Distributional System Compatibility	Dangerous for Prolonged Exposure (ppm)	Environmental/Safety Properties
Distillate Oils	Mix	19,000	494	150	22	Diesel compatible	Distribution system in place	500	...
Ethanol	C_2H_5OH	11,930	793	235	30	Manifold heating required Cold start problem below 58°F	Water solubility in ethanol causes special handling requirements	1,000	...
Gasoline[b]	Mix	19,290	430	145	20	Compatible	Compatible	500	...
Hydrogen[b]	H_2	51,620	1,085	200	105	Requires conversion	Leakage hazard Materials deterioration problem	Nontoxic[c]	Safety hazard
Methanol	CH_3OH	9,080	878	280	41	Manifold heating required Cold start problem	Water solubility in methanol causes special handling requirements	200	Lower NO_x emissions High aldehyde concentration in exhaust
Methane[b]	CH_4	21,250	1,170	165	45	Compatible	Compatible	Nontoxic[c]	Safety hazard
Propane (LPG)	C_3H_8	19,940	808	180	27	Compatible	Compatible	Nontoxic[c]	Safety hazard

a Energy equivalent of 20 gallons of gasoline.
b Cryogenic liquid.
c Asphyxiant.

Table 4 Problem Areas of Alternative Fuel Choices[8]

Fuel	Vehicle Tankage[a]	Engine Compatibility	Distribution System Compatibility	Synthesis Process Technology	Resource Availability	Toxicity
Distillate Oils	✓	D[b]	✓	✓	✓[d]	✓
Ethanol	✓	✓[c]	D	✓	D	✓
Gasoline	✓	✓	✓	✓	✓[d]	✓
Hydrogen	D[b]	✓[c]	DD[b]	✓	✓	✓
Methanol	✓	✓	D	✓	✓[d]	✓
Methane (SNG)	D	✓	D	✓	✓[d]	✓
Propane (SLPG)	✓	✓	D	DD[b]	✓[d]	✓

[a] No great difficulty—✓; some difficulty—D; serious difficulty—DD.
[b] At present, may not present problems in the far term.
[c] If blended up to 10% with gasoline.
[d] At present, may present problems in the far term.

gaseous forms. Conversion, distribution, and in-vehicle storage characteristics are strongly dependent on their physical state. For example, utilization of gaseous fuels (hydrogen, methane) poses in-vehicle storage problems. If stored at ambient temperature and pressure, enormous storage volumes are required relative to that needed for the storage of an equivalent energy value of gasoline. If the gas is compressed, significant tank weight penalties are incurred; and if the gas is liquefied, again weight penalties are incurred, plus refrigeration is required to maintain cryogenic temperature. Hydrogen storage as metal hydrides alleviates the cryogenic problem but suffers from high cost and weight penalties. There are, however, tradeoffs between these cost penalties and processing costs to produce other fuels that are more compatible with existing systems. Table 4 indicates specific problem areas for the different fuels.

The technological processes available and under development for the production of the previously listed alternative fuels are shown in Figures 8-12. The accompanying tables (Tables 5-9) briefly summarize current development status, perceived technical risks and potential barriers to commercialization, estimated date of initial commercial availability, conversion plant efficiency, economic data, and environmental/sociopolitical considerations.

Crude oil refining (Figure 8) is an established technology, and the problems or barriers that exist for future development of the petroleum system (Table 5) result from the economic, environmental, and sociopolitical problems inherent in developing for production frontier areas, particularly Alaska and the offshore continental shelf, and the high costs of enhanced recovery from existing reservoirs. Production of transportation fuels from coal gasification (Figure 9 and Table 6) and liquefaction (Figure 10 and Table 7) has the advantage of utilizing the extensive domestic coal resource base, although cost breakthroughs are needed. Processes that can produce motor gasoline at reasonable costs should have high priority. Shale oil (Figure 11 and Table 8) offers the potential of contributing significantly to domestic liquid fuel supplies in this century. In situ processing offers substantial potential environmental advantages over surface retorting, and could be the key to overcoming barriers that limit the ultimate contribution of shale oil to domestic supplies. Biomass conversion could provide a truly inexhaustible resource if the required fuel forms can be produced. There are economic problems associated with large-scale collection of thinly distributed agricultural waste and final product distribution to retail outlets, and there is a requirement for basic research on processes and study of nonprocess constraints such as land and water use. Also there exists a social

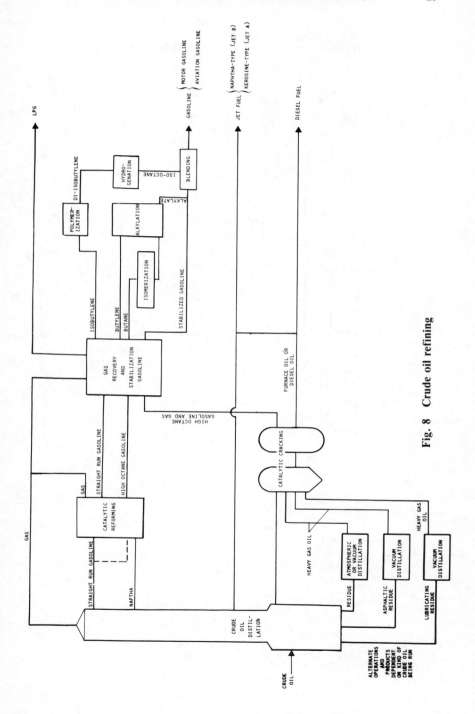

Fig. 8 Crude oil refining

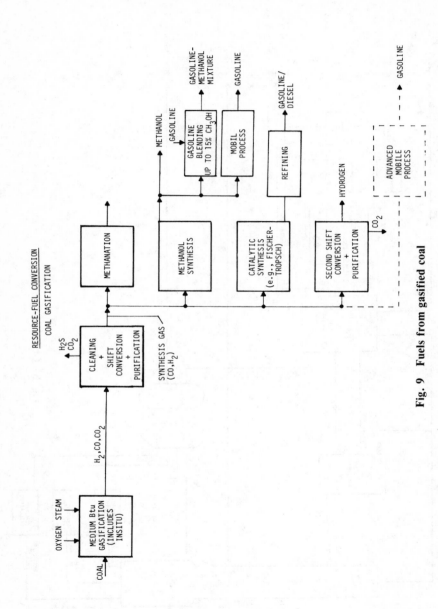

Fig. 9 Fuels from gasified coal

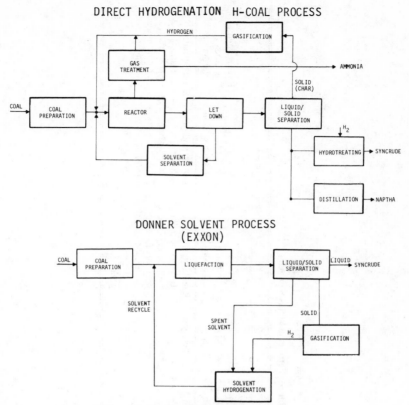

DIRECT HYDROGENATION H-COAL PROCESS

DONNER SOLVENT PROCESS
(EXXON)

Fig. 10   Fuels from liquefied coal

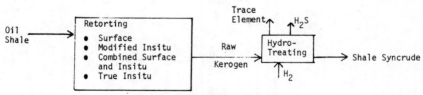

Fig. 11   Oil shale conversion

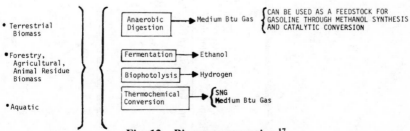

Fig. 12   Biomass conversion[17]

Table 5   Petroleum Status Summary [11-13]

Resource	Technical Parameters				Economic Parameters		
	Stage of Development	Development Risk/Barrier	Estimated Date of Initial Commercial Availability	Plant Efficiency	Fuel Cost ($/10^6 Btu)	Plant Capitalization	Environmental/Social Parameters
Crude Oil	Commercial	None for current operations. High costs of future Alaskan, offshore, and enhanced recovery operations.	Current	88%-96%	Regular gasoline at full-service retail outlets, 1977 average purchase price for dealers, 54.2¢/gallon, including 12.5¢/gallon tax.	350-500 million dollars for a 200,000 B/D U.S. grass-roots refinery, 1976 completion.	Future offshore operations may be constrained by environmental considerations.

Table 6 Fuels from Gasified Coal: Status Summary [14-16]

Fuel Production Path (by Fuel Type)	Technical Parameters			Process Efficiency	Economic Parameters		Environmental/ Social Parameters
	Stage of Development	Development Risk/Barrier	Date When Ready for Commercialization		Estimated Fuel Cost ($/$10^6$ Btu)	Plant Capitalization	
Substitute Natural Gas	1st generation gasification technology commercialized. 2nd generation nearing demonstration stage.	No major barriers in 1st generation.	1st generation available. 2nd generation 1983-1985.	1st generation 59%	1st generation 4.5 (1975$). 2nd generation 3.6 (1975$).	$1B for 84 trillion Btu/yr plant (1975$).	Water availability. Impact of large processing facility construction/operation.
Methanol	1st generation gasification technology commercialized. Methanol synthesis.	Improved process needed.	1st generation 1980. 2nd generation 1983.	1st generation 52%-56%	7.20 (1975$).	$500 M for 32 trillion Btu/yr plant (1975$).	Same as above.
Gasoline (from Methanol)	1st generation gasification technology commercialized. Mobil process: successful pilot plant; demonstration planned.	Moderate risk. No major technical barriers foreseen.	1983	45%-50%	...	...	Same as above.
Gasoline Diesel	Fischer-Tropsch technology demonstrated (e.g., SASOL). 2nd generation processes under development.	None for 1st generation processes.	1st generation available.	35%-60%	...	...	Same as above.
Hydrogen	Emphasis on methane generation. Hydrogen processes in early pilot plant stage.	Moderate; processes still not fully developed.	1981	...	...	...	Same as above. Materials problems.

Table 7   Fuels from Liquefied Coal: Status Summary [15]

Fuel Production Path (by Fuel Type)	Technical Parameters				Economic Parameters		Environmental/ Social Parameters
	Stage of Development	Development Risk/Barrier	Estimated Date of Initial Commercial Availability	Plant Efficiency	Fuel Cost ($/10^6 Btu)	Plant Capitalization	
Syncrude	Small pilot plant operational.	Moderate-high risk.	1987 (for both H-coal and Donner-solvent processes)	55%-80% (H-coal total energy conversion)	Syncrude: 4.7 (1975$).	$1.28 for 130 trillion Btu/yr plant (1975$)	Impact of large processing facility.
Naphtha	Large pilot plant in construction.	Major barriers: catalyst recovery; liquid product separation.			Naphtha: 4.7 (1975$).		Large water requirements.

Table 8 Oil Shale Status Summary[15,16]

Process Types	Technical Parameters				Economic Parameters		Environmental/ Social Parameters
	Stage of Development	Development Risk/Barrier	Estimated Date of Initial Commercial Availability	Plant Efficiency	Fuel Cost[a] ($/$10^6$ Btu)	Plant Capitalization	
Surface Retorting	Retorting technology in pilot plant phase. Technology for upgrading shale oil (after retorting) available from petroleum processing has recently been demonstrated by SOHIO, DOE, and DOD.	Moderate risk. Major barriers: solids handling; mechanical difficulties with commercial-sized retorts.	Available	70%-80%	3.0 (1975$)	$980M for 83 trillion Btu/yr (1975$)	State and federal emission controls may limit production. Aquifer disruption from mining. Leaching from spent shale piles. Severe reclamation problems. Impact of large mining/processing operation. High water requirements.
Modified in situ Retorting	Feasibility-testing phase.	Moderate to high risk. Major barriers: mine development; improved fracturing techniques required; enhanced retorting efficiency. Diversity of resource requires different processes in order to operate in different geologic formations.	Available	50%-70%	2.7 (1975$)	$850M for 100 trillion Btu/yr plant (1975$)	Subsidence. Aquifer contamination. Some pollution from leakage through the overburden.

[a] Primary product; does not include additional processing required to produce transportation fuels.

Table 9　Biomass Status Summary [10,15,17]

Process	Technical Parameters		Estimated Date of Initial Commercial Availability	Conversion Efficiency	Economic Parameters		Environmental/ Social Parameters
	Stage of Development	Development Risk/Barriers			Product Cost	Plant Capitalization	
Anaerobic Digestion	Pilot plant testing complete. Demonstration plant in planning stage.	Small risk. Major barriers: operational sensitivity; corrosion; scale-up.	1982	...	...	$19M for 1.3 trillion Btu/yr plant (1975$)	Facilitate future disposal of municipal wastes. Possible material salvage. By-product disposal problem.
Fermentation	Feasibility testing. Pilot plant to start mid-1980. Similar processes currently commercialized in Brazil.	Small risk. Major barriers: control of hydrolysis kinetics; enzyme production; enzyme reuse.	1990	...	...	...	...
Biophotolysis	Conceptual stage.	Medium-high risk. Major barriers not identified.	...	...	...	...	...
Thermochemical Conversion	Feasibility testing. Pilot plant to start construction beginning 1981.	...	1990	Electricity 30-35%. Methanol 42-48%. Ethanol 42-48%.	Electricity 24 mils/kWh. Methanol $0.36/gal. Ethanol $1.27/gal	...	...

problem in diverting agricultural products to fuel use when much of the world's growing population is undernourished.

The fossil and biomass resources listed previously can, of course, also be burned in powerplants to produce electricity to power electric vehicles. Electricity can be produced in numerous other ways as well. In fact, the DOE Inexhaustible Energy Resources Planning Study[10] notes that the advanced inexhaustible energy technologies now under development are oriented toward producing significantly more electricity than is expected to be demanded and significantly less nonelectric energy than demanded.

In the transportation sector, electricity has in the past been used to power trains, subways, and other vehicles for which electrical current can be conveniently supplied externally. This has limited its application for detached vehicles which would require onboard energy storage. Consequently, most electric vehicle R&D is oriented toward improving storage capacity. Electric vehicles offer the potential, particularly for urban and suburban transportation, of reducing the dependence on petroleum products with a vehicle that is environmentally benign. Electricity at a user source has a low delivered efficiency (about 30%) and a high price (about $12/MBtu based on 4¢/kWh) relative to a gasoline efficiency of about 90% and a current pump price of about $5/MBtu. However, the gasoline motor is a far less efficient converter of fuel (22% to 28% efficiency) than is the electric motor (45% to 58% efficiency). The net effect is that the cost per mile of electricity is somewhat greater than that of gasoline. Figure 13 provides an energy flow diagram for electricity utilization, and Table 10 provides operating cost comparisons for a range of battery types. The MOPPS study[6] concluded that the electric and hybrid vehicles are unlikely to achieve significant market penetration until the latter part of this century.

8.   The Government's Role

The technology development programs initiated by the former U.S. Energy Research and Development Administration (ERDA) and taken over or initiated by the U.S. Department of Energy (DOE) are designed to demonstrate technical, environmental, social, and economic feasibility of new technologies in order to accelerate their adoption by the private sector. This is being accomplished through a program of grants, cost-shared contracts, loan guarantees, and other assistance.

The ultimate success of the RD&D function of the Department of Energy will be the degree to which these new technology options prove commercially successful in meeting or reducing future energy demands. Previously under ERDA,[21] the strategy for accomplishing this objective for each technology area consisted of (a) identifying and

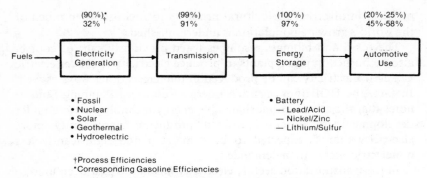

†Process Efficiencies
*Corresponding Gasoline Efficiencies

Fig. 13   Energy flow diagram for electricity utilization [18]

prioritizing needed RD&D efforts and the appropriate government role, (b) maintaining a supporting research program in government and academic laboratories to supplement industry efforts and provide a scientific basis for improved technology, (c) establishing and implementing an orderly development program culminating with demonstration projects, usually cost-shared with private industry, and (d) providing for documentation and dissemination of data to implement technology transfer throughout the private sector. Currently under DOE, the strategy is considerably more oriented toward early commercialization of the most promising technologies. The inclusion into DOE of mechanisms for implementing loan guarantees and other

Table 10   Comparison of Energy Storage and Costs for Vehicles [19,20]

| Power Source | Specific Energy (W-hr/lb) | | Cost (1973 $/mile)[b] | Comments |
	Base	Adjusted[a]		
Gasoline	1130	140	0.13-0.15	Current state-of-the-art
Lead/Acid Battery	20	15	0.18-0.25	Near term; 54 miles nominal range
Nickel/Zinc Battery	50	35	0.20	Near term; 144 miles nominal range
Lithium/Sulfur Battery (Design Goal)	140	100	0.14-0.15	Long-term goal; 145 miles nominal range

[a] Adjustment of base number to take account of the efficiency advantages of an electric motor over a heat engine.
[b] Based on electricity at $0.036/kWh and gasoline at $0.50-$1.00/gal (1973 dollars).

incentives allows programs to be structured more toward achievement of market penetration, rather than just demonstration of feasibility. Commercialization of a new technology will take place only if it is available at a cost that allows the private sector an acceptable rate of return on the capital required. As noted earlier in this paper, this is a necessary but not sufficient condition for commercialization. Various institutional, environmental, and sociopolitical barriers must be overcome, and the Department of Energy has programs underway to alleviate some of these problems.

While commercial viability of a new technology requires bringing it to the point where the above condition can be met, a government-supported RD&D program may terminate somewhat short of this point and a variety of government incentives and/or regulations may provide the final impetus. These include loan guarantees, price supports, direct procurement of products, special tax credits, and regulations or penalties for undesirable energy usage. The government can judiciously supply these incentives and regulations to help create a market, which should result in the cost savings expected to accrue from economies of scale, and ultimately these incentives may be unnecessary, as conventional fuels deplete and costs rise.

This is conceptually illustrated in Figures 14 and 15. Figure 14 schematically illustrates a technology development curve, showing that funds must be invested to make cost reductions. The total investment can be provided by both government and the private sector.

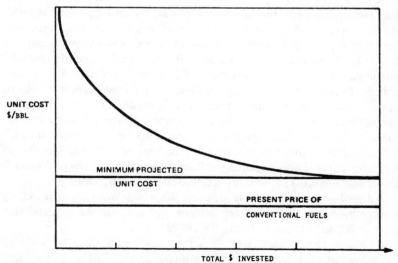

Fig. 14 Technology development curve[17]

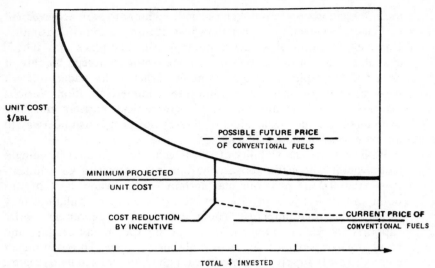

TOTAL $ INVESTED

Fig. 15   Effect of incentives and conventional fuel price rise[17]

It is clear that in some instances the minimum projected unit cost remains above the existing price of conventional fuels. In this case, there will be no demand for the technology sought to be introduced. This does not imply that the technology will be too expensive for all time and therefore not worth pursuing.

Figure 15 shows that cost to the consumer can be reduced by incentive. The objective is to reduce costs of a new technology through R&D to a point near, but not necessarily at, the cost-competitive level. At that point, the situation can be improved by two circumstances. The first is that the cost of conventional energy sources may continue to rise in the future. If it rises above the minimum reducible cost of a particular technology, a demand pull is created on the basis of cost competitiveness. The other circumstance is that cost reductions can be made by various incentive devices. Tax credits or other incentive mechanisms can effectively drive the price into the fully competitive range, and the technology products can thus be "commercialized." These incentive mechanisms may be reduced or removed in time, as learning curve and economics of scale productivity improvements lower end-use costs, or as conventional fuel costs increase.

Tables 11-13 indicate those technologies which would be affected by government incentives and regulations and indicate the approximate sizes of RD&D programs in the FY 19798 budget.

In other federal activity, Congress has established a National Transportation Policy Study Commission, to "make a full and complete investigation and study of the transportation needs and of

Table 11   Transportation Technologies with Federal Role Indicated—Highway Mode[6]

Market Sector (Technology)	Federal Actions Assumed in "Current Role"				
	DOE Program (Yes/No)	Millions of R&D$ FY 78	Total	Incentives (Yes/No)	Regulations (Yes/No)
Light-Duty Vehicles					
Waste Heat Utilization	Yes	0.50	10	No	No
Chassis/Body	Yes	3	50	No	No
Heat Engine Propulsion					
Continuous Combustion Engine	Yes	29	460	No	No
Continuous Variable Transmission	Yes	1	10	Yes	No
Accessory Drive	Yes	0.30	2	No	No
Intermediate Trucks					
Waste Heat Utilization	Yes	a	a	No	No
Chassis/Body	No	a	a	No	Yes
Heat Engine Propulsion	Yes	a	a	No	No
Large Trucks					
Waste Heat Utilization	Yes	1	15	No	No
Chassis/Body	No	...	3	No	Yes
Heat Engine Propulsion	Yes	10	10	No	No
Small and Medium Cars					
Electric and Hybrid Vehicles	Yes	46.90	311	Yes	No
Multiple					
Highway Vehicle Refrigeration	Yes	0.05	11	Yes	Yes
Driver Conservation Training	Yes	0.25	160	Yes	Yes
Van Pools	No	0.05	...	Yes	No
Car Pools	No	0.05	...	Yes	No
Emergency Fuels— Alternative Fuels	Yes	0.25	0.5	No	No
Emergency Diesel Fuel Dilution	Yes	0.10	2	No	No
New Hydrocarbon Fuels	Yes	0.50	20	No	No
Alcohols	Yes	1.30	19	No	No
Hydrogen and Other Long-Range Fuels	Yes	0.25	13.7	Yes	No
Synthetic Gasoline/ Diesel Fuels	Yes	0.20	6	No	No

[a] Current programs are costed in other vehicle classes.

the resources, requirements, and policies of the United States to meet such expected needs." Assessment of energy requirements and availability of energy to meet anticipated needs is an integral part of the Commission's charter. A final report will be submitted to the President and the Congress no later than December 11, 1978.

Table 12 Transportation Technologies with Federal Role Indicated—Aviation and Pipeline Mode[6]

	Federal Actions Assumed in "Current Role"				
Technology	Doe Program (Yes/No)	Millions of R&D $ FY78	Total	Incentives (Yes/No)	Regulations (Yes/No)
Air Mode					
Engine and Airframe Design—Airline	No	88	670	No	No
System Design and Operation—Airline	No	0	0	No	No
System Design and Operation—General Aviation	No	0	0	Yes	Yes
Pipeline Mode					
Pipeline Bottoming Cycle	Yes	0.90	10	No	No
Turbocompounding Reciprocating Engines	Yes	0.20	5	No	No
Friction Reduction Techniques	Yes	0.20	2	No	No
Alternative Fuels	Yes	0.10	20	No	No
Control System Optimization	Yes	0.20	2	No	No

Table 13 Transportation Technologies with Federal Role Indicated—Rail and Marine Modes

	Federal Actions Assumed in "Current Role"				
Technology	DOE Program (Yes/No)	Millions of R&D $ FY 78	Total	Incentives (Yes/No)	Regulations (Yes/No)
Rail Mode					
Propulsion Systems	Yes	0.35	60	No	No
System/Component Design Improvements	Yes	0.08	40	No	No
System Operating Procedures	Yes	0.04	30	No	No
Marine Mode					
Powerplants and Auxilaries	Yes	0.25	50	Yes	Yes
Propulsors	Yes	0.20	25	Yes	No
Hydrodynamics	Yes	0.02	25	Yes	No
Operational Improvements	No	0	0	No	No
Structural Improvements	No	0	0	No	No

9. Summary

The key points made in this paper are summarized below:

(a) Annual energy consumption in the transportation sector has been projected to grow from the current level of slightly less than 20 Quads to roughly 38 Quads in 2010. Successful market penetration of conservation technologies could reduce this demand in 2010 to about 25 Quads.

(b) Automobiles are currently the largest consumer of transportation energy and, consequently, savings in automobile energy consumption are probably the key to a more efficient transportation system.

(c) Most of the savings potentials occur through implementation of conservation technologies in highway vehicles, mainly deriving from improvements in heat engine propulsion systems.

(d) Despite these savings, the resulting demand for transportation energy, coupled with declining domestic petroleum reserves and the problems associated with increasing dependence on crude oil imports, highlights the need for alternative fuels.

(e) Competitive prices will not alone ensure widespread usage of a new fuel. Institutional barriers, particularly those resulting from the inflexibility of the national transportation system infrastructure, must be overcome to achieve commercialization.

(f) Currently, gasoline is by far the transportation fuel most heavily demanded in the U.S. today. However, a shift toward middle distillates is occurring. By the year 2000, demands for diesel and jet fuel may each exceed the demand for gasoline.

(g) Alternative transportation fuel usage (e.g., electricity, alcohols, hydrogen) will be relatively small through the year 2000. Consumption of gasoline and middle distillates is expected to dominate the market through the rest of this century.

(h) Fuels that can be produced from fossil resources and from biomass include distillate oils, ethanol, gasoline, hydrogen, methanol, methane, and propane.

(i) In terms of R&D for alternative liquid fuels, liquids from biomass could provide a truly inexhaustible supply if the required fuel forms can be produced. Priority should be given to production of conventional distillace-like fuels compatible with conventional liquid fuel requirements. However, production of alcohols may be of increasing value as blends with gasoline or an direct-use fuels, subject to the introduction of engines that can utilize alcohol fuels. In addition, alcohol fuels used in stationary power sources can free other fuels for use in the transportation sector.

(j) High priority in coal liquefaction research should be given to processes that can produce motor gasoline at reasonable cost. The potential for obtaining alcohols from indirect liquefaction processes needs further study, particularly in support of a transportation system modified to utilize alcohol fuels.[22]

(k) The Department of Energy is supporting the development, demonstration, and commercialization of alternative transportation fuels. In addition to supporting research through grants, cost-sharing contracts, etc., financial incentives can and will be utilized to accelerate the penetration of new technologies into the marketplace.

Comments on Adams' Paper

(a)   One of the sociological factors that should be considered in the planning of major changes in transportation-fuel supply is the differing time constants for major changes in lifestyle. For example, a house in the suburbs might have a half-life of 50 years whereas the automobile that would carry the suburban commuter to the city and back would have have a half-life of 10 years. Therefore, automobile engineering could turn over many times before there would be any major change in where people live. Accurate predictions of such time constants would be helpful for the engineer, but were best supplied by the social scientists.

(b)   It might be advisable to set transportation performance standards at miles-per-energy-unit, e.g., miles per million Btu rather than miles per gallon. This would provide a consistent measure to cover a wider cut of fuels than just gasoline.

Another helpful figure would be the net energy yield of a particular fuel calculated across the system. For example, if for gasoline the goal is to save energy by getting 27.5 miles per gallon, and this is attained not only by building smaller and lighter cars, but by increasing the engine compression ratio, it would call for more severe refining to get to a higher octane. The commensurate additional energy expenditure at the refinery should be taken into account in any measure of net energy saved to get to the 27.5 miles per gallon.

B.   Summaries of Prepared Remarks and Commentary

1.   John P. Longwell

The major end-products of our hydrocarbon supply, the ones for which we seek alternative sources, are gasoline, jet and diesel fuel, the lower-grade distillates, heavy fuel oil, and chemicals. (The demands of the chemical industry—which are not within the province of this Symposium—are nevertheless expected to be an increasingly significant drain on hydrocarbon sources. For example, the lighter cars indicated for the future will in all likelihood contain a large portion of plastics derived from petroleum or its equivalent.) Among the considered alternative sources are shale oil, coal, and biomass. The properties of the sources determine to a great extent their respective end-products. Shale oil, for example, is so much better than coal for jet and diesel fuel that even though it is fairly aromatic one would avoid using it for making gasoline. Coal liquids, on the other hand, are uniquely qualified for gasoline because they are mostly aromatics, or can be converted very readily to aromatics by normal refining. They are a poor source for jet and diesel fuels and for olefins; and they are a good source for fuel oil and low-grade distillates. Char or biomass can be converted to carbon monoxide (CO) and hydrogen and are therefore sources of methanol, good quality gasoline, hydrogen, very high-grade diesel fuel, and jet fuel. (Biomass is also a prime source of chemicals for plastics.) Since this refining process is expensive, one would not use it to make lower-grade products. The difference in cost per Btu of going the CO-hydrogen route as against, say, coal liquefaction to high-grade products such as jet and diesel fuel, is small enough at this point not to be a factor for consideration.

The changing ratio of gasoline to mid-distillates mentioned by Adams has major implications. We will be changing from an era in which we had a surplus of paraffins, and were able to make all the good quality diesel and jet fuel we wanted, to an era of surplus in aromatics. Aromatics are very poor for diesel and jet fuel, but excellent for gasoline. Since petroleum, shale oil, and direct coal liquids all contain aromatics which can be made into gasoline, the possibility exists that a very high-aromatic gasoline can be a desirable alternative fuel, because such a fuel can do some very special things. Specifically, if we look at the motor octane numbers of the aromatics, we see that they are extremely high. We could perhaps go up high enough in engine compression ratio to compete directly with the diesel engine and thus help offset the diminishing output of diesel fuel. The use of highly aromatic hydrocarbons would call for as big or bigger changes in automobile design as, say, shifting to methanol, but a consideration of the relative advantages and disadvantages would seem to be indicated. To look at it more broadly: if we examine the opportunities to reoptimize the whole system—that is, aim for maximum benefits for the least amount of additional refining energy—we may be led to hydrocarbon fuels as one of the major alternative fuels for the future.

2. Jack Grobman

Jet fuel price increases and projected jet fuel shortfalls due to shifts in future supply and demand have led to serious considerations of the actions necessary to prevent a constraint on the future growth of air transportation. These actions must be addressed to the overall conservation of energy in both the air transportation and petroleum refining industries. NASA, along with other government agencies and private industry, has been conducting a research and technology effort to establish the data base necessary to optimize future jet fuel characteristics in terms of refinery energy consumption and tradeoffs in jet aircraft and engine design complexity. Other research and technology efforts are being conducted by NASA to reduce jet fuel consumption by improving aircraft energy efficiency by reducing specific fuel consumption, reducing engine weight, reducing aircraft weight, and improving aircraft aerodynamics.

Broadening current jet fuel specifications would permit reductions in energy consumption at the refinery. A broad-specification fuel may be defined arbitrarily as a liquid hydrocarbon fuel with key properties or characteristics that substantially exceed current specification limits for aviation turbine fuels. As for alternatives to the dwindling supply of petroleum, the prime candidates, shale oil and coal syncrudes, will not be available for aviation in substantial amounts until after the turn of the century. Cryogenic hydrogen, with its present high costs and the major changes it would require in airframe and ground handling systems, also appears to be a long way off. So for the next 25 to 30 years, we will be depending very heavily on petroleum as a source of aviation turbine fuels.

Jet fuel has traditionally been manufactured by distillation from petroleum crude followed by a mild hydrogen treatment to control sulfur, corrosivity, or thermal stability as needed. Crude petroleum normally has a boiling range that extends to about 600°C. As the demand for jet fuel, diesel oil, and heating oil increases, a point will be reached where there is an insufficient quantity of material in the proper boiling range. It will then become necessary to convert fractions boiling above 300°C to these lower boiling products. These cracked products are, in general, higher in aromatic content (increased carbon to hydrogen ratio) than are the naturally occurring fractions. The processing required to produce current specification jet fuel from the higher boiling fractions consumes considerably more energy because of the process hydrogen requirements than does the conventional production of jet fuel by crude distillation.

The commercial jet aircraft fuel, Jet A, has a relatively narrow boiling range specification. The initial boiling point, a minimum of about 170°C, is necessary to keep the flash point above 40°C to reduce the probability of a fire during fueling or following an emergency landing. The final boiling point for Jet A is usually below 270°C to comply with limits on the freezing point. Freezing point increases as the final boiling point is increased. The freezing point of a fuel blend is the temperature at which wax components in the fuel begin to solidify. The specification for Jet A limits the freezing point to a maximum of $-40°C$.

The average aromatic content of Jet A has increased from 14% (by volume) in 1960 to about 17% in 1976. The ASTM Jet A specification for aromatic content is a maximum of 20%. During the emergency period of 1973-1974, the limit was temporarily raised to 25%; more recently, a waiver has permitted the limited use of Jet A with a maximum aromatic content of 25%. During the emergency period, the limited quantities of Jet A refined from heavy Arabian crude had aromatic contents as high as 22%. Projections indicate that Jet A refined from Alaskan crude may have aromatic contents as high as 25%. These increases in aromatic content may be attributed in part to the production of Jet A by distillation of crudes with relatively higher aromatic content. In the future, several other factors could cause the aromatic content of jet fuel to increase. As mentioned earlier, the cracking of higher boiling materials to produce a product within the jet fuel boiling range increases the fuel aromatic content. The aromatic content of jet fuel may also be increased by extending the distillation range to a higher final boiling point.

In the future, as the relative demand for jet fuel increases, it will be necessary for refineries to consume considerable quantities of hydrogen in order to meet the requirements for current specification aviation turbine fuel. Since the production of hydrogen requires significant energy consumption, and since hydrogen and the processes using it are very expensive, consideration of cost and energy conservation encourages minimizing these types of refining. Thus, there is a definite need to investigate the effects of broadening jet aircraft fuel specifications on jet engine performance and durability in order to develop a data base which will allow an optimization of future fuel characteristics that takes both refinery energy consumption and aircraft engine design tradeoffs into account. In order to implement this optimization effort it is desirable to establish a target fuel for use in research programs on both fuel production and aircraft/engine design.) A NASA-sponsored workshop, June 1977, has recommended a fuel for experimental use with the following characteristics:

	Jet A	Experimental fuel
Hydrogen, wt.%	~14	~13
Aromatics, vol.%	<25	~35
Flash point, °C	>40	>40
Freezing point, °C	−40	−29
Breakpoint temperature, °C	>260	>240

Comparison of the representative values for the properties of the proposed future broad-specification fuel with those of current Jet A fuel indicates that the major changes to be expected would be (a) an increased aromatic content corresponding to a reduction in hydrogen content, (b) a higher final boiling point, (c) a higher freezing point, and (d) a lower thermal stability JFTOT (jet fuel thermal oxidation test) breakpoint temperature. The thermal stability breakpoint temperature is an empirical laboratory indication of the degree to which the fuel may be heated without incurring significant levels of fuel decomposition. The properties designated for the future broad-specification fuel tend to be similar to those of the current number 2 diesel fuels. Increases in aromatic content will result in increased smoke and flame radiation. A higher final boiling point will likely result in a less volatile, more viscous fuel, which will affect both ignition characteristics and idle emissions, and a higher freezing point, which will affect the pumpability of the fuel.

The problems that are most likely to be encountered as a result of these modifications relate to engine performance, component durability and maintenance, and aircraft fuel-system performance. The effect on engine performance will be associated with changes in specific fuel consumption, ignition at relight limits, and exhaust emissions. Durability and maintenance will be affected by increases in combustor liner temperatures, carbon deposition, gum formation in fuel nozzles, and erosion and corrosion of turbine blades and vanes. Aircraft fuel-system performance will be affected by increased deposits in fuel-system heat exchangers and changes in the pumpability and flowability of the fuel. In short, if we anticipate a broader specification fuel, then we face a critical need for research and technology to enable the aircraft to handle it.

Counteracting adverse fuel property effects on engine performance, such as reductions in ignition and relight limits and increases in exhaust emission levels, will require advanced combustor technology such as improved or auxiliary fuel atomizers, better control of fuel-air mixing and distribution, and lean combustion techniques. Counteracting problems related to component durability and maintenance will require such advanced technology as improved fuel atomizers, lean combustion techniques, thermal-barrier coatings, and new materials. Solving problems in aircraft fuel systems will require fuel-tank heating techniques and "tailored" elastomer materials. Even though preliminary evaluations of several of these technological advances have been encouraging, considerable research and development is still needed to make them acceptable in production engines and aircraft fuel systems. Furthermore, the ability to cope with several other problems, such as those caused by variations in thermal stability and by trace constituents, has not been demonstrated to even an acceptable experimental level at the present time. The factors that contribute to variations in thermal and chemical stability are not well understood and much more research is needed. Turbine erosion and corrosion problems may be somewhat relieved by using coatings, but considerable research is needed to fully understand all the factors that contribute to these problems.

Because of the many unknowns that must still be explored and explained through research and development efforts, it is apparent that these efforts should proceed at an orderly and timely pace. Although it is not certain that aircraft will have to operate with the wide variation in fuel properties discussed herein, a sound and complete technological data base must be developed as soon as possible if the aircraft industry is to have any impact on setting acceptable variations in the specifications of future aircraft fuels. It is none too soon to start developing this data base since tradeoffs will have to be made to determine the optimum choice between the cost and difficulty of developing advanced engine and fuel-system technology and the economic advantages to be gained by reducing the degree of refining needed to produce current-specification fuels from projected future fuel feedstocks.

3.   Robert Rightmire

There is general agreement that total systems must be considered when deciding among the various alternative transportation fuels of the future. Simply stated, the total system approach here is to take all the hydrocarbon resources, decide what is being gone after (which fuels, which chemicals), set the constraints (environmental and otherwise), and then get to work. The point to keep in mind as we devise the total system is that we should use the minimum amount of processing necessary to satisfy the objective. In other words, the less one has to manipulate the hydrocarbon molecules—or, put another way, the less hydrogen required in processing—the lower the costs, and the better the system.

To illustrate this in a working example, let us assume a set of conditions for the year 2000, based on our best predictions. (This example is somewhat simplified for discussion purposes, since it is based on a liquid-hydrocarbon system and does not include natural gas or biomass.) Our analysis has told us that the end-products demand in the year 2000 will require resources amounting to approximately 23 million barrels per day of liquid hydrocarbons, up from about the 18 million we use today. This is substantially less than was forecast as little as two years ago, because of the impact of conservation efforts both in automotive fuel economy and in the growing movement toward coal for industrial and utilities boilers. The 23 million estimate (see Figure 16) was reached as follows:

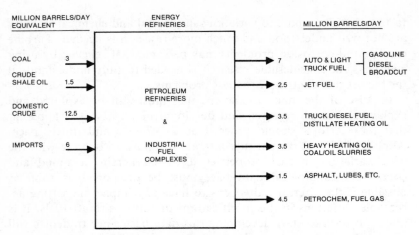

Fig. 16   Liquid hydrocarbon supply-demand: year 2000

Automobile and light-truck fuels—7 million barrels per day. This figure is essentially the same as it is today, but today it consists for the most part of gasoline. However, inroads will increasingly be made by the diesel engine and the fuel-injected stratfied-charge engine which can operate on a broad-cut fuel. Our forecast looks for a substantial drop in consumption by the late 80's or early 90's from today's 7 million barrels, but then we expect it will rise again—not so much from automobile use, but from increasing light-duty truck, light-van, and recreational-vehicle use.

Jet fuels—2.5 million barrels per day. This is up from today's 1 to 1.5 million. We do not see the product greatly changing in quality from today for a variety of reasons, not the least of which is safety.

Heavy diesel fuels—over 3 million barrels per day. This figure also includes distillate heating oil, and is a threefold increase from today's 1 million. The reasoning was clearly laid out by Adams.

Heavy heating oils, marine diesel fuels, bunker fuels, etc.—3.5 million barrels per day. Even though we predict a slight increase in demand, we are keeping the figure at today's 3.5 million in the expectation that the increase can be met by coal oil slurries and coal char slurries.

Asphalt—1.5 million barrels per day.

Petrochemicals—3 to 4 million barrels per day. This is the biggest jump, up from today's less than 1 million, with the biggest growth coming between 1985 and 2000.

Fuel gas—0.5 million barrels per day.

Now let us look at the forecast for the supply side. From Alaska, with luck, we should bring in 3 million barrels of crude oil per day. In the lower 48 states, using today's techniques, we could expect a drop to as little as 4 million barrels of crude. With enhanced recovery and discovery techniques, we might get up to about 7 million. From offshore, we could look for a maximum of 2 to 3 million. The total then of domestic crude in 2000 would be a maximum of 12.5 million barrels per day. The forecasts for *imported* crude are highly uncertain. A peak of 70 million barrels per day in world oil production is predicted for the years 1985 to 1990, up from today's 60 million, followed by a drop and a substantial world demand and price increase. Therefore we should not depend on more than 6 million barrels per day in the year 2000, or a total crude oil supply, domestic and imported, of 18.5 million.

The challenge of the system, then, is to fill the gap between the 23-million-barrel demand and the 18.5-million-barrel supply with coal derivatives and crude shale oil, and to do it as simply as possible in order to minimize what would be a tremendous economic impact.

Let us assume that we want to fill 3 million barrels of the gap with coal derivatives. Many coals, such as Illinois #6 seam and Pittsburgh #8 seam, contain hydrocarbons that can easily be converted to a liquid stream and a char in roughly 50% yields: for our example, 1.5 million barrels of liquid, plus the oil equivalent of 1.5 million barrels in char. Refineries at present, just to provide heat, consume about 1 million barrels per day. In our example, the simplest thing to do would be to take all the char and burn it for refinery heat. As for the other 1.5 million barrels, new fluid-bed combustors are being developed to work in actual processing technologies. When they come to pass, half of what's left might be used there. The remaining half could be taken up in the area of coal-oil slurries, where a good deal of testing is now going on. Thus, relatively simply, i.e., without having to put in a lot of hydrogen, we have been able to move the coal.

Now let us look at a way of handling the crude shale oil: If we split it down the middle, taking half in a distillable product and half in a residual product, the boiling point is about 750°F at the splitting point. If we can make diesels operate on a 750° end-point fuel, then we can take the top half of our product, and without coking or hydrocracking, make it into a heavy diesel fuel—reducing greatly the hydrogen input in the processing. One of the problems we would have to solve in order to make this work is the smoke problem; we would have to learn how to atomize and spray these heavy materials into diesel engines. But that kind of research would pay heavy dividends. The point here is that, by looking at the total system, we can design the engines to operate on what we can produce, rather than designing the product for existing engines, with the expensive hydrotreating that implies.

What has been outlined is one approach to the alternative fuel dilemma. There are many possible approaches, and many scenarios that might be constructed. The emphasis in this one is simplicity. The simple answer may indeed be the best answer.

4. Alternative Transportation Fuels—Goals and Objectives

Jon B. Pangborn
Institute of Gas Technology

Our hopes and plans for alternative fuels are predicated on several notable goals:

(a) To utilize domestic natural resources in an efficient manner.

(b) To thereby reduce imports of foreign oil and aid our international balance-of-payments situation.

(c) To produce fuels that are reasonably economic (to the consumer) relative to new increments of conventional domestic fuels and to other alternatives.

(d) To provide fuels that are physiologically and environmentally safe and convenient in delivery, storage, and utilization, and which, if possible, are compatible with existing fuel systems.

We should keep these goals in clear focus when specific program objectives are decided and when associated projects are carried out. Another possible goal, subject to debate, is to accomplish the above as soon as possible.

Debate on this point is inevitable, because the time frame impacts economics, technologies, and hence fuels and fuel systems.

Consistent with the above goals are several general tasks:

(a) We must consider complete fuel systems, that is, from the process of resource extraction or conversion through utilization by the consumer.

(b) We must decide which fuels meet which technical market requirements in the most economic manner over a span of time.

(c) Until we are committed to development of a given fuel system, we must assess and periodically reassess, through engineering studies, the technology and economics to produce, deliver, and utilize promising candidate fuels.

(d) When we commit to a fuel system, we must know why, when, where, how, and for how long.

Currently, we seem to have a "mixed bag" of efforts on alternative fuels for transportation. Our focus on alternative fuels for certain markets seems fuzzy, and our efforts seem divided.

For example, the aircraft industry first desires no alternatives; rather, commercial airlines would like to continue with crude-oil-based distillate fuels for the indefinite future—"somebody else should change." (The airlines, of course, are not alone in holding this opinion; all market sectors seem agreed on this point.) After the wish for no change, distillate fuels from oil shale or coal seem preferred. Yet, Lockheed Aircraft has made a convincing case for hydrogen as an aircraft fuel based on overall system economics and efficiency. After natural gas and oil as feedstocks, hydrogen could most economically be made, at large scale, from coal. Should we be considering the hydrogen-from-coal option more intensively?

Another example of conflicting opinions occurs in the choice of alcohols, gasoline from alcohols, or diesel fuels from coal for automotive transportation. Agricultural interests and some elected representatives are pushing for alcohol (methanol and/or ethanol) supplements to gasoline, and for the blend "Gasohol" ("Gasahol" in Illinois), while some extend their preference to a pure alcohol automotive fuel. Recently, however, we observe General Motors to be pushing further the commercialization of diesel engines for automobiles. On the basis of production and delivery, both the Institute for Gas Technology (IGT) and Exxon (Esso) Research and Engineering, in their original studies on alternative fuels for automotive transportation (for EPA, circa 1974), concluded independently that distillates (including diesel oil) from coal and oil shale were preferred technically and economically in the near- (until 1985) and midterm (1985-2000) futures. Methanol and ethanol, however, can be produced now, with existing technology—including methanol from coal—if it's really needed. Recently, Mobil Oil has provided a technology for making gasoline from methanol. What is our objective for alternative automotive fuels? Do we want very near-term capabilities and an agricultural-based fuel (ethanol, gasohol), or mid- and far-term capabilities of large scale (distillates from oil shale or distillates and methanol from coal)? Until we have a decision, we will be in the mire of assessments, reassessments, and re-reassessments.

Finally, a remark is in order about our progress with "paper" studies versus chemistry, engineering, laboratory work, field testing, and pilot planting. I cite alcohol as a convenient example to make the point, and I quote from an abstract:

> "In alcohol-gasoline blends, either the alcohol used must be anhydrous or blending agents must be added. In engine performance, the fuel consumption for the same power and acceleration will be greater than with gasoline. Benzene and toluene have been found to be very satisfactory blending agents. Further work might prove that the higher alcohols are equally suitable. Absorption of water in the stored fuel must be avoided. Engine starting with the mixed fuel is more difficult than with gasoline of the same A.S.T.M. 10% point. The octane rating may be raised from 2 to 10 points by use of 10% alcohol in the fuel."

The article of this abstract was written by O. C. Bridgeman and published in *Industrial Engineering Chemistry*, News Edition, Volume 11, pp. 139-140, in 1933. Thus, our technical utilization problem with alcohol was made pretty clear by Bridgeman 45 years ago.

Currently, the only way to get further with alternative transportation fuels is to know our goals, follow through with the task work so that we can commit to practical alternative fuel systems, and then do the experimental production, storage, delivery, and utilization tests in a connected, coordinated, engineering project. Realistic funding levels for this will be in the tens of millions, but to do nothing will eventually cost the consumer tens of billions.

5. Symposium Participants

(a) The projected import level of 6 million barrels per day is based on very optimistic predictions for the domestic product. If they do not come to pass, the import requirement could go as high as 13 million. Since that is a prohibitive figure, it is doubly necessary to find a simple solution and get it on the books as quickly as possible. As we move beyond the year 2000, and the oil begins tapering down, other options will begin taking over. The gradual increase in the cost of energy will allow these options to become economically viable. Included here are biomass systems and solar energy, especially as they can be utilized to produce the less expensive fuels.

(b) It was suggested in the MOPPS study that the market factor will determine the course to be taken in alternative fuels, and since world oil production is expected to peak by about 1985, the timing of the changes to alternative fuels may be considerably sooner than we are talking about. This is a difficult and controversial area to nail down, since the conclusions must be based on assumptions, and the assumptions are being drawn from a huge body of studies and reports that do not necessarily agree. For example, what one ends up believing about the role of the USSR is pivotal. Whether they continue as an exporter of crude oil or instead become an importer will affect the market hugely. Another example is the differing assumptions one can pull out of the MOPPS data on fossil-based resources. If one goes to the high side, the date prediction for market penetration of alternative technologies is anywhere from 1995 to as far as 2010. If one goes to the low side, significant market penetration begins in 1985. In any case, it is highly unlikely that "all of a sudden, in 1985, on the 32nd day of July, at 4 in the afternoon, all the world is going to run out of oil"—nor, as has sometimes been postulated, will there be a precipitous decline over some three-year period. The system does not work that way.

(c) It is generally agreed that we must establish priorities as we approach the entire problem of alternative fuels. And yet with all of the studies, we are still in a situation where DOE's budget on liquid fuels is but 1.5% of their total budget, and where we have still not done anything about studying the fuels and combustors together as a total system. How can we go about setting priorities when there are basic needs that we have not yet uncovered?

(d) In 1976, ERDA made a decision to stockpile a petroleum reserve for emergency purposes. The decision won out over a competing idea—to put the money into a large synfuel plant to supply the emergency fuel. The reasoning was as follows: given a fuel emergency of three months or longer in the near term—1976 to 1981, while the stockpile would still be relatively low—money spent on the synfuel plant would buy greater benefits. In the middle term—1981 to 1984—it would be a toss-up. Beyond 1984, by which time there would be a year's supply in the stockpile, benefits from the stockpile would outweigh those from the plant. And during that critical year, other arrangements could be made if they had not already been. The decision came down to what one believed about the security of our supplies in the near term versus the far term, and the thinking was that the far term was shakier—that dependence on imports beyond 1984 might be considerable.

Nevertheless, a case can be made for a commitment to one type of technology that seems practical by taking it through the planning and demonstration stage. This would strengthen our near-term position; it would hold lessons for us in technology and economics; it would be a small step toward easing our balance-of-payments problem; and it still would allow us to pursue other alternatives in other directions.

C. Workshop Session Discussion

1. Matching Fuels to Applications

In view of the enormous complexities associated with energy supply aspects of alternative transportation fuels—a plethora of potential fuels, the possibilities for engine modifications, the effects of economics (both classic and unpredictable), environmental impact, social costs and benefits—we must be very careful not to set the kind of rigid priorities that can shut off particular lines of development.

We must also be aware of the tendency to look almost exclusively at near-term and middle-term solutions to the fuel problems in our various efforts to protect our own industrial interests. It is understandable that a given segment of the transportation industry might feel its need for a continued supply of its present fuel so vital that the most efficient use of alternative fuels would be for everyone else to switch—thereby releasing the supply of standard fuels to where it obviously could do the most good: to themselves. These problems are after all very real and very serious—in some cases to the point of survival. Certain of the transportation industries—aircraft in particular—must commit themselves now to new families of engines that will be around for the next 35 years.

The decisions prioritizing who gets to use what fuels for the near and middle terms come down, finally, to government regulations. But for the DOE to set intelligent priorities, it needs a great deal of data, both technical and economic. And perhaps the most useful kinds of technical data for them to start with are long-term data: our assessments as to what the long-term solutions, the post-petroleum solutions, are likely to be. From that kind of perspective we can work backward and find it easier to set priorities for the near and middle term. At the same time, it is up to the transportation people to present and explain to the Congress and to the DOE transportation's special needs—because if they don't, nobody will. Or, to put it another way, nobody *can*—because nobody knows them so well.

The basic point that must be made is that transportation, in order to be effective, needs that fuel which contains the maximum amount of energy in the minimum amount of space. While the logic of that is unassailable, it somehow gets thwarted by political considerations. And every constituency gets its political protection. As a result, we find ourselves expending great amounts of personal energy figuring out ways to convert coal to the equivalent of petroleum; whereas it perhaps would be more efficient to use some of that kind of energy to convince the rule-makers how much more sensible it would be to let the powerplants and the home heating units burn the coal since they have a reasonable amount of space for it, and give petroleum to transportation. This would extend the supply of petroleum, and buy us many more years in which to develop viable alternatives.

2. The Market As a Development Driver

It may very well be that the solutions for industry lie in economics, that is, in realistic economic predictions based on fuel pricing determined by the market, rather than by politicans succumbing to pressures to hold fuel prices down. Given this kind of situation— which perhaps is too much to expect in the real world of politics— DOE economists could reasonably predict various fuel prices for at least the next twenty years. This possibly would be the very best thing that the DOE could do for industry *and* the consumer. With this kind of information, industry could commit itself to spending money now on developing alternative fuels for the future, and for adapting their hardware and distribution systems to it, because it would have the data to determine cost-effectiveness. It may even be that the aviation industry, for example, might decide that in that open fuel market, it will be able to afford to stay with petroleum-type fuels—including fuel from shale and coal—that have climbed in price to as much as fifty dollars per barrel.

However, if we wish to let the marketplace decide on the alternative fuels, we have to develop engines that can accommodate a broad range of specifications. Two areas in particular lend themselves to this kind of broadening: automobile/light truck, and heavy truck. And these account for half of the oil that's going to be used. If the broadening of engine capabilities can go hand in hand with refining modifications, the change to alternatives can be made gradual—which is greatly to be desired. It would be very risky to be forced to make a sudden jump from, say, an engine that runs on 50 cetane to one that will run on 20. It would take tremendous planning and money; beyond that, the success of the entire system would be very vulnerable all along the line to any slip-up in timing.

It would help those of us who are trying to tie our decisions to economics, and who are trying to get some money together to develop alternative fuel options, if we had some absolute numbers: each option's cost estimations and thermal efficiencies, arrived at with great care on the same basis, so that they can be accurately compared. Granted that costs on imaginary systems can be estimated only up to a point—a point beyond which one would have to build and find out. Nevertheless, loose numbers still abound; one can look for any kind of number he needs to fit his preconception and find it. The effect in the marketplace is to undermine people who really want to get something done, but find themselves unable to compete with other people's numbers.

3. Timeliness

The one fact whose impact pervades this entire discussion, that indeed called this Symposium into being, is that the time for decision-making is *now*. If we wait very much longer, that is, until we have full information on all of the many alternatives we will be discussing, it may simply be too late. The world fuel situation is at a point where OPEC can close the valve just a couple of turns and we will have a major depression in this country; where the USSR has greater oil reserves than we do; where the USSR and Cuba are making inroads in North Africa. We must get some plants going for some alternative fuels, and quickly.

There are various points of view on how to go about getting something done. The most aggressive of them is simply to go ahead and do it. To wit: nobody has all the answers in advance, and if you wait for them all you'll be left behind. It is too easy to put off doing anything at all while waiting for the results of another economic study, or for the development of an improved process. Better to get on with it: decide what looks best based on present information, build the plant, and incorporate any future improvements as you go along.

The attitude from Washington on getting something done—and from the Congress in particular—is that time is quickly running out; the pressure of the international economic system is forcing a reaction. If decisions and actions do not come in time from the technical community, then they will come from the policy makers, based on whatever technical information they have. The key point is that the time for decisions on energy will be determined not by how long it takes the energy people to complete their studies and analyses, but by much larger issues. The technical community can influence these decisions or not, depending on when it acts.

4. The Role of Government

Obviously the government role in energy decisions is formidable, not only in the regulations and laws that it *will* establish, but in the ones that it already has: for example, price control. The energy industry, by and large, would vastly prefer that the government get out of energy pricing, and allow the give and take of economics to determine prices. Given that circumstance, they could go ahead with reasonable risk and invest in plants for alternative fuels. But without that circumstance, the risks inherent in having to outguess the Congress are *un*reasonable. The other side of the issue, the view of it from the Congress and the agencies, is that they arrive at their laws and regulations from a broader frame of reference. Energy policy is one component, albeit a vital one, in the matrix of government policy. It interacts, for example, with employment, with international economics, with whether the poor will get some heat in a freezing winter.

The energy establishment can acknowledge the need for a Congressional role in energy decisions, but feels strongly that the proper exercising of that role must not include dictating which technology to use. An example of constructive Congressional action is the Energy Policy and Conservation Act which set 27.5 miles per gallon by 1985, but without spelling out how to achieve it. A similar forcing function could be implemented with fuels. For illustration (and these are only illustrations, and in no way recommendations), a law might state that, beginning in 1983, all new cars, at the point of sale, have their tanks filled with a nonpetroleum fuel. Period. That would be enough to support say a 10,000-barrels-per-day plant. It would be a start. Or a law might require that, by a given date, 1% of all automobile fuel be synthetic. There are any number of approaches that Congress can use to force development and provide incentive for building alternative-fuel plants—so long as it does not say what those plants should be.

Lest this be construed, despite disclaimers, as a plea for regulations, one must add that there are dangers here that must be thought through. The specter of fifteen different energy companies all coming up with disparate products portends a situation of such economic complexity as to suggest a deeper hole for the industry and the country than the one they are in now.

It might be useful to look at current government interaction (other than R & D) with alternative energy. The only thing that passes for a major alternative energy policy across the land is the advocacy of coal. Everyone is now waiting for EPA's interpretation of the best available control technology on the use of coal, and this unintentionally will affect to a great extent where the coal will come from and the amounts produced. It could serve to drop the bottom out of the low-sulfur coal market in the northern Great Plains. If, on the other hand, we just allowed the market to work out what coal would be produced—based on productivity, man-hours per Btu, etc.—this would jeopardize a lot of new technologies that are waiting in the wings for cleaning up coal throughout the cycle. Thus, there is a question as to whether national energy-supply strategy should go into these kinds of details at all, and to what extent they should be considered as incentives in order to get some legislation passed. For the present there is no direction from the Congress—and it can be argued that what industry *really* needs are simply loan guarantees or other iron-bound incentives. This would get some real production going.

Yet another attitude, that of many industry people toward government involvement, is that it should be limited to proof of concept and development of the technology. From that point on it would be industry's decision to go with it or not, as it sees fit.

Government's role, however, cannot be separated from the public's attitudes. When the public gets concerned enough, it pushes the legislature, state or federal. California's Proposition 13 attests to that. In terms of energy problems, be they shortages or pollution, the same thing is happening; or so it is in California. Other states and the federal government will be forced to take notice. By early 1979, California, in response to public pressure, will have a fleet bill calling for a large number of cars to be purchased by the state government and running totally on nonpetroleum fuel—coal or biomass— produced under a state-mandated program.

5. *Incentives*

While they are waiting for their legislatures to move, people are moving on their own. For example, in home heating, which is probably the energy-use sector most flexible in its choice of sources, a lot of individuals are burning wood in wood stoves—even though that idea may be appalling to engineers. But there is a gulf, when problems become problems for individuals, between what they do and what conclusions the so-called leadership is arriving at during its con- siderations of global engineering and economics.

Another incentive by way of legislation is coming out of California. A bill, now before the Assembly, after being carried in the Senate by 27 – 3, gives to the distributor a tax credit of one dollar per gallon of alcohol made from California agricultural products or waste, when he uses it as an additive to gasoline. The proportion can be whatever he chooses. Originally the bill called for a tax credit on the blend itself at five cents per gallon, provided the blend contained at least 5% alcohol. The present bill, as altered, gives greater flexibility to the distributor in his use of the alcohol, and also simplifies the tax-accounting procedures for the state. A companion bill now before the California Assembly spells out a policy in which the state will have in place, at the end of ten years, a prototype synthetic-fuel industry, based on coal and biomass. On the state level it is easier to move with these initiatives than on the federal level—and state legislators in California, after waiting in vain for some leadership and direction from Washington, are moving ahead. Perhaps this kind of local action will press DOE and the Congress into becoming more decisive. Since it may take up to forty years of lead time to get us self-sufficient in energy, the time to start is now.

But it should be pointed out, in fairness to the Congress, that national political realities are just as real as industrial realities; and the political reality here is that Congress is a reactive institution. Given some kind of positive commitment from industry, the Congress will respond positively. As an example, a small chemical company in Kansas sent some people to Washington to press for legislation. For two years this company had worked on a process of making ethylene not from petroleum but from corn cobs. They would now be ready to go ahead with it except that the economics were not right. (This, incidentally, is something the giant chemical companies do not want to get into at all.) Because of this commitment to a process by one small company, and because of the company's willingness to act aggressively in Washington, it is very likely that a tax-incentive measure will get through the Senate to encourage production of industrial chemicals from biomass.

To some, however, this example smacks of excessive government planning of the economy. Hundreds of ideas, they say, are presented aggressively to the Congress and there are at least as many bad ones as there are good ones. In the final analysis, it is the market which is the most effective instrument for allocating resources along the most efficient lines (for evidence, see the USSR experience of the last fifty years). Government strategy may properly include the use of a social discount rate, rather than industry's private discount rate, in order to capture future social costs; but it should still be designed to allow the marketplace to do the job of separating the good from the bad.

6. Government/Industry Interaction

It is evident that, for the industry-government symbiosis to work, each needs something from the other. Industry is looking for guidance from Congress. Congress is looking for some commitment from industry. It would therefore be very helpful if these two sets of minds could come together and become as one. Toward that end, *effective* representatives of each should sit down together, figure out what it is that each other needs, and pledge to go after it. Out of such a meeting would come some genuine movement.

A simplified but perhaps useful restatement of industry's point of view vis-a-vis government is this: there is a national energy problem which is affecting our international balance of payments and the value of our currency, and even, possibly, our security. To combat the problem, we need right now to develop alternative fuels. Industry can come up with the technology and the fuels, but not without losing money. It is prepared to lose money to solve its own problems; but the solution of *national* problems should be paid for from the national fund. If, on the other hand, the situation were not as described, if indeed the development and production of alternative fuels showed reasonable promise of being a profit-making enterprise, there would be no symposia such as this one. Industry would be doing what industries always have done in our economic system: investing its resources in order to earn a profit.

When we extend this simplistic model of the problem, we find ourselves again in the morass of price control. If the federal government would allow fuels to be sold at the world market price, that would be all the incentive industry would need. It could then assess its risks based on reasonable predictions and jump into the areas of alternative fuels from solid ground. But it is impossible to jump when the ground keeps shifting under you. The government's idea of incentive seems to be to provide tax relief in the form of x number of dollars per barrel of synthetic fuel produced. And while this helps set some kind of monetary level on what these fuels might be worth, it is still artificial. The question as to whether it is enough still remains. A more productive exercise of the government's control capability would be to apply it to the amount of oil we will import over the long term, i.e., lay out a specific import schedule over the next ten to fifteen years. This would guarantee a market to industry. Then, if that were coupled with decontrol of energy prices, industry would go about the business of providing alternative fuels.

7. Keeping the Options Open

Any apparent emphasis on a particular alternative fuel thus far in this discussion should not be taken as a technological preference of this Symposium. It is simply illustrative. Preferences have not yet been

arrived at—indeed, preferences are bound to vary, depending on the
job to be done. The feeling throughout, sometimes voiced and
sometimes not, is that there are three sources alternative to petroleum:
coal, shale, and biomass; and we are far from ready to rule any out.
We must move ahead on all of them as best we can. Only then will we
find out what works and what doesn't. It also seems clear that many
solutions, even to the same energy problem, are likely to emerge, with
no one alternative source emerging as a clear "winner."

Alternatives cannot be measured on any basis of equivalency
because they cannot ever be equivalent. The forces driving the
development of each alternative—economic groups, companies,
societal considerations—are disparate enough to dispel any possibility
of "all things being equal."

IV.
STORAGE AND DISTRIBUTION

A. Storage and Distribution of Cryogenic Fuels

W. E. Timmcke
Air Products and Chemicals, Inc.
Allentown, Pennsylvania

1. Abstract

The cryogenic fuels LNG and LH_2 may be considered as alternatives to the fuels presently utilized in the vehicles that provide our transportation. Should these alternatives prove to be attractive, the technology available to the designer of storage and distribution equipment for these cryogens will play an important part in the development of the logistic support system which will be required. The present state of the art of storage and distribution equipment and its applications to the various transportation modes is discussed. It is concluded that the capability to store and transport these potential fuels is well within the designers expertise and that the evolution of a total support system for a specific transportation vehicle will improve skills already developed.

2. Introduction

Cryogenics is the branch of physics concerned with phenomena that occur at very low temperature. This discussion will deal with only two of these cryogens, LNG and LH_2, which can be considered as having a potential use as alternate fuel for transportation equipment. It is not my purpose to discuss the production or utilization of these two cryogenic fuels, but rather to limit the discussion to the storage and transportation equipment required to develop a network for the logistic support of the various transportation modes which might become dependent on these cryogenic fuels.

Storage and distribution of the cryogenic fuels, liquid hydrogen and LNG, is today a reasonably common everyday occurrence. However, until a market develops for widespread use of these fuels, the industries that provide the storage and distribution equipment and service can only stand by, prepared to meet the developing market demands. For the purpose of this presentation, we are going to assume that the cryogenic fuels can be economically produced, which will, in turn, spur the imagination of the various vehicle and engine designers to proceed with the development of the vehicle whether it be the automobile, locomotive, or airplane.

This discussion will deal with the storage structures and distribution containers designed for those products, liquid hydrogen and LNG, whose normal boiling points at atmospheric pressure of -423&F and -259&F, respectively, fall within the cryogenic range. Since 1957, cryogenic storage and distribution has become part of industrial economy primarily as a result of expanding the use of cryogenic liquefied gases used in the space program and then the gradual application of this experience to the metallurgical, electronic, food processing, and chemical industries.

First let us consider the storage facility requirements at the fuel-producing location and at the using activity, be it an airport tank farm or individual filling station for our personal automobile. The industry that builds large-volume cryogenic storage facilities has the background and experience to thoroughly understand the design of foundations, materials of construction, vessel configuration, insulation, fittings and appurtenances, cleaning requirements, and connecting facilities.

3. Configuration

Storage vessels for cryogenic fuels have been built in many sizes ranging from those that would handle liter quantities in one-foot-diameter dewars for laboratory use up to more than one-million-gallon tanks, nearly 200 feet in diameter. In general, cryogenic liquids are stored in double-walled vessels. The inner and outer walls of the container are normally concentric and of the same basic configuration. Several basic shapes have evolved as the requirements for cryogenic vessels have expanded, and these are classified based primarily on the shape of the inner vessel.

Flat-bottom cylindrical tanks may be used for vapor space pressures ranging from slightly over atmosopheric to up to five pounds per square inch gage. The roof closures are usually ellipsoidal or spherical with the bottoms normally flat, although sometimes the tank bottoms are constructed with a shape similar to that of the roof of the tank. The inner container rests upon some form of load-bearing insulation which carries the weight of the stored fluid to the foundation. The bottom of the tank serves only as a seal and is not subject to significant stress. However, provisions must be made to hold down the inner shell against lifting forces arising from internal pressure. Hold-down devices must permit the vessel to move readily in response to thermal displacements. Heating elements beneath the load-bearing bottom of the tank must also be provided to eliminate freezing of the moisture in the subsoil and subsequent heaving.

The flat-bottom design provides cryogenic storage at minimum cost, but other shapes may be dictated by pressure transfer, heat leak, or foundation requirements. To date, flat-bottom tanks have been used for LNG storage, and, although some design work has been done on this configuration for liquid hydrogen storage, all large vessels currently in use for liquid hydrogen are spherical containers.

Spherical containers have been built utilizing the basic perlite insulation system for LNG, or, if lower loss rates are required as in the liquid hydrogen case, the vacuum-perlite systems can be used. Spherical configurations usually become most economical when pressure storage or vacuum insulation is required and the tank capacity is in the range above 30,000 gallons. The soil conditions at the storage site often dictate the consideration of spherical vessels rather than flat-bottom tanks such as when soil conditions dicate the use of piling to support the vessel.

Cylindrical tanks with end closures are designed for internal pressure and are adopted for either the vacuum-powder insulation system or the so-called super insulation system, depending on the tolerable loss or boil-off rate. Cylindrical tanks satisfy many of the same conditions as the sphere, but present some economic advantage. Its smaller size allows complete shop fabrication, shipment to the site, and installation as a complete unit. As might be expected, the smaller the vessel, the greater the boil-off as a percent of rated capacity because the surface area does not decrease as rapidly as the volume decreases.

It should be recognized that all of the above types of tanks are within the designers' capabilities and, with the selection of the proper insulation techniques, the boil-off rate can be controlled to rates of 0.5% per day to as low as 0.1% per day for liquid hydrogen and to even lower rates for the higher-boiling-point LNG.

4. Materials of Construction

One of the most important considerations in the design and construction of vessels for cryogenic service is the selection of the proper material. How the material will perform at cryogenic temperatures depends on such properties as notch toughness, ductility, critical flaw size, specific heat, coefficient of thermal expansion, and conductivity, as well as the usual strength and elasticity. Considerable research has been done in recognition of the problems associated with proper material selection and fabrication techniques so that industry is prepared to meet requirements for cryogenic storage.

While most metals increase in strength with a decrease in temperature, some, such as carbon steel, almost completely lose their ductility at cryogenic temperatures, making them unsuitable for the construction of the inner vessels. Copper, nickel, aluminum, and most alloys of these metals exhibit little ductile to brittle transition and therefore are suitable for cryogenic service. Without going into the details of specific construction materials, it should be explained that selection of materials is well within the designers' capability. Such construction meets minimum requirements established by the ASME Code, API Standard, and publications of regulatory bodies governing construction of vessels for ultralow temperatures.

5.   Design Conditions

Once the materials of construction and the specific functions are determined, the design procedure for cryogenic vessels is not unlike that used for conventional tanks. The geometric volume below maximum liquid level and required minimum vapor space must be determined. Vapor space determination is based on the operating conditions that the tank is required to meet. In most cases where the vessel is associated with the production plant, the product that is flashed in the storage or transfer system can be recycled through the processing plant. This eliminates the losses which would result from venting the normal pressure buildup.

The inner vessel for LNG storage must be designed to accommodate internal pressure arising from vapor-phase pressure and liquid weight. Since product density for LNG can vary based upon the source of the gas, particular attention must be given to the highest density product in determining tank design. For vessels using the perlite insulation system, the inner vessel design must take into consideration the external pressure from the insulation, insulation space gas pressure, and a small allowable internal vacuum. In the case of the vacuum-powder insulated vessels, the internal vessel design pressure must include allowance for reduced external pressure due to the vacuum between inner and outer shells. Special consideration must also be given to concentrated loads in the region of support attachments and fittings, taking into account thermal displacements. Suspending the inner tank within the outer shell demands that the loading and thermal changes be carefully considered, with sufficient distance between inner and outer attachment to minimize the heat conducted through the insulation space.

Outer vessel design loads include wind, snow, dead loads, internal pressure from insulating powder, and insulation space gas pressure. When vacuum-powder insulation systems are used, the outer vessel must be capable of meeting all the normal outer vessel design loads plus the external pressure of 15 psi.

6.   Insulation

All of the storage vessels for cryogenic fuel require some type of insulation. The basic insulation for large-volume storage vessels is composed of a granular material produced from a special volcanic rock known as perlite. The perlite is contained in the space between the inner vessel which contains the product and the outer vessel which acts as a gas tight and weather protective barrier. This barrier retains the perlite and protects the insulation from external forces such as rain, wind, humidity, fire, and flying objects. Normally, the perlite is expanded in an on-site portable furnace and placed in the insulation space while hot. This method minimizes moisture in the insulation which must be kept dry by a purge gas system to a dew point lower than the storage temperature. In a vacuum-powder insulation system, moisture in the perlite would prevent a satisfactory vacuum from being maintained. Filling the insulation space utilizing the on-site portable furnace also minimizes breakdown of the perlite particles since they are handled only once after expansion.

Higher quality insulation than the basic system for LNG and the vacuum-powder system for liquid would seldom be required for large storage containers. However, if additional insulation capability is needed, the annular space may be evacuated to a very low pressure to reduce gas conduction and convection to negligible levels. If even greater insulation capability is required, the so-called super insulation system can be used. This consists of layered reflective insulations wrapped around the inner tank to reduce heat transfer by radiation and convection so that when the annular space is evacuated the heat leak is reduced to levels considerably below those of the vacuum-powder systems. While this technique has not proved economical for large containers where boil-off can usually be recovered or safely vented, we will see the importance of this form of insulation as we proceed to consider transport vehicles and user storage requirements.

7. Transport Technology

The technology for transporting large quantities of cryogenic fuel is available if anybody needs it. Tankers, railcars, and barges are all used within the United States to transport the cryogenic fuel from the production or storage point to the ultimate user. It should be pointed out that, with the exception of the liquid hydrogen used by the Space Program, the primary reason today for production of LH_2 or LNG is the economies of transportation and storage.

The materials of construction and insulation required by the designer of cryogenic transport vehicles are similar to those previously discussed. The new problems that confront the designer are the physical size or envelope dimensions allowed, and the dynamic forces on the moving vehicle. LNG tankers today are in the size range of 10,000 to 13,000 gallons. Only minor increases in that size are possible due to the axle loading reaching highway limitations, or the physical size exceeding dimensional limitations. These trailers are used to serve satellite storage of LNG terminals or peak shaving facilities.

Liquid hydrogen trailers in today's service have a capacity of approximately 13,000 gallons. Due to the low density of the liquid hydrogen, the vehicle weight and axle loading do not present problems; however, trailer size, the 40-foot length, 8-foot width, and 12.5-foot height, limit the usable capacity. Some improvement could be achieved by utilizing the maximum allowable dimensions, but the manufacturing economics of building a tank trailer of a configuration other than the current cylinder within a cylinder prevents this method of increasing the payload from receiving much attention.

Railcars for liquid hydrogen transportation are currently in service hauling quantities of 35,000 gallons each. These cars also are reaching the size limitation of the rail system. With dimensions of nearly 90 feet long, 10 feet wide, and 15 feet high, they become some of the largest cars capable of being moved on the United States rail system. Cars of this size require particular attention in routing since curves in excess of 12 degrees, having a turning radius of approximately 475 feet, limit available roadway or siding facilities.

As mentioned previously, the super insulation technique is most often used in the manufacture of transport vehicles where the limitations are size rather than weight. The vacuum space between the outer and inner walls of the container is minimized, thus maximizing the usable cargo volume. Using this insulation technique also holds the boil-off rate to less than 0.5% per day. This is very important in over-the-road transport of liquid hydrogen as the regulations established by the Department of Transportation do not allow venting of the product between filling the tanker and its arrival at the user's off-loading location. With the limited number of production locations and the widespread locations of users, this is a critical design consideration. With the advent of more production points and thus shorter hauling distances, this nonventing requirement may become less critical.

While moving liquid hydrogen over the road, the driver/operator must be constantly in attendance with the vehicle. For that reason trips in excess of 10 hours require a driving team. Railcars have a unique feature in this respect since they obviously travel great distances in unattended condition; therefore the boil-off rate must be kept to a minimum and provisions must be made to safely vent the container should internal pressure build up in excess of the regulated amount. For this type of situation, an automatic mixing device is incorporated in the vent system which mixes the venting hydrogen with air to assure that the percent of hydrogen in the vented gas stays below the flammable limits. A system of this type may become a practical solution to the venting problems that may occur should cryogenic fuels become a possibility in our future transportation system.

8. Pipeline Distribution

Before leaving the subject of distribution equipment, some mention should be made of the potential of pipeline distribution of cryogenic liquid fuel. The NASA space program at Cape Kennedy provides the only example of today's practice regarding pipeline distribution of liquid hydrogen. In the fuel system for the Space Shuttle, approximately 1500 feet of vacuum-jacketed line comprise the fuel transfer system which will be used to transfer 400,000 pounds of liquid hydrogen for each flight. The fueling operation is brief and is scheduled to take place every nine days at maximum launch schedule. In an infrequent fueling operation such as this, no provisions are made for recovering the boil-off losses which can amount to close to 25% of the actual fuel used. A system to transfer liquid hydrogen in small quantities would not prove to be economically feasible. However, should a situation develop where large quantities are required in close proximity of the liquefier, and a boil-off return line can be incorporated to reprocess the vent gases, pipelines would deserve some consideration. This would be the case where a liquefaction plant was installed adjacent to a major airport.

Storage at the user locations presents no unusual problem and suppliers of cryogenic gases are well aware of the criteria for matching user requirements to supplier capabilities. While few ground storage vessels provide for liquid withdrawal, as most customers use the product as a gas, liquid withdrawal and transfer over short distances can be accomplished within existing technology. Tank sizes up to 30,000 gallons can be shop fabricated and installed as complete units. Sizes in excess of this can be field-erected without difficulty.

9. Future Projections

In considering a transportation system based on a cryogenic fuel production and distribution system, some thought must be given to the user's mobile storage container. Currently there are three methods available, or at least being considered, for the mobile storage of hydrogen for a future automobile:

(a) Compressed gas.
(b) Liquid hydrogen.
(c) Metal hydrides.

In evaluating each technology we must be aware of the following parameters:

(a) Total system weight—the maximum stored energy density per weight.

(b) Total volume—the maximum energy density per volume must be obtained.

(c) System cost minimization.

(d) Safety and ease of operation.

(e) Maintenance.

Many techniques for storing hydrogen fail to satisfy these points. Most of the problems associated with storing hydrogen can be attributed to a relatively low energy density (Btu/ft^3). For example, the energy density of gaseous hydrogen at one atmosphere is only 29% that of natural gas. On the other hand, liquid hydrogen exhibits a value which is equivalent to 28% to that of gasoline. Table 14 compares several forms of hydrogen storage with typical gasoline and other fuels.

Compressed hydrogen provides a mode of storage that has some beneficial aspects such as long shelf life and simple control systems, but this method suffers from two problems: (a) the need for a large number of cylinders to contain reasonable quantities of fuel, and (b) the large system weight associated with storage cylinders. As an example, a cylinder having a water volume of 1.5 cubic feet and a weight of 160 pounds contains only slightly more than one pound of hydrogen at a pressure of 2400 psi. Current research on advanced lightweight hydrogen cylinders suggests that the total system weight for compressed gas storage may be reduced 40-50%. While this could have significant impact on fixed storage and distribution equipment design, the projected weight reductions are not sufficient to meet the requirements of mobile storage since a reasonable quantity of fuel storage could still exceed the overall vehicle weight.

Hydrogen can also be stored by chemically combining it with various metals as a hydride. With the application of heat, the hydride is dissociated and the hydrogen gas is released for use. Hydrogen densities larger than liquid hydrogen densities can be obtained by this

Table 14   Characteristics of Cryogenic and Other Fuels

Fuel	Density (lb/ft^3)	Energy Density Btu/lb	Energy Density Btu/ft^3	Amount Equivalent to 20 Gallons Gasoline Gallons	Amount Equivalent to 20 Gallons Gasoline Lb
Typical Gasoline	43.8	19,000	832,000	20	117
Gaseous Hydrogen (2400 psi)	0.787	52,000	41,000	406	43
Liquid Hydrogen (NBP)	4.42	52,000	230,000	72	43
Methane Gas (1500 psi)	5.26	22,000	116,000	143	100
Methane Liquid (NBP)	26.5	22,000	583,000	29	103
Hydrogen in Metal Hydrides					
FeTiH$_2$	362	993	360,000	46.5	2,180
Mg$_2$NiH$_4$	159	1,900	302,000	55	1,140
MgG$_2$	108	3,970	429,000	39	550

method and, therefore, it is currently generating strong interest. The main drawback at the present time for mobile applications is the very high overall weight and volume of such systems. Further research is warranted for the development of lightweight, inexpensive materials which absorb large quantities of hydrogen and readily liberate hydrogen at near-ambient temperatures and pressures.

It is premature at this time to estimate the potential economics of a metal alloy hydride system, but the metal alloy, storage facility, reactor vessel, heat exchanger, and control systems of a hydride system equivalent to 20 gallons of gasoline would probably cost in the neighborhood of $1000 based on the present state of the art. However, if we are planning to place hydrogen in the hands of the general public in a mobile storage system, the reduced safety hazard in this form of storage may be well worth the price.

Under certain conditions hydrogen released from a storage vessel can auto-ignite, but when stored as a metal hydride it is considerably less active and hazardous. There is some evidence that upon ignition a metal hydride has a propensity to self-estinguish. This characteristic is attributed to the rapid formation of an oxidized metal surface which acts as a barrier for further hydrogen release. Various metal hydrides, when fully saturated with hydrogen, have been shown to be insensitive to impact, shock, ignition, and explosion. Hydrogen released by the rupture of a hydride vessel will be minimal since the sudden pressure drop will promote endothermic desorption of the hydrogen, rapidly reducing the bed temperature, causing the subsequent rate of hydrogen desorption to drop to very low values. The introduction of oxygen into the vessel promotes oxidation of the metal surface, causing the desorption rate of the hydrogen to drop to lower values. This could prove to be a desirable safety feature should a metal hydride mobile storage system be chosen for the vehicles of the future.

10.   Summary

It is fairly evident from today's technology that the cryogenic fuels are not going to be transported as liquids over great distances by pipeline due to high heat leak and economics. Therefore, if cryogenic fuels are to be used for transportation, it will be necessary to consider a different type of production and distribution system for each transportation mode. In the case of the large quantities required for fueling aircraft, the economics would require liquefaction of the gas at a site adjacent to the airport with the liquid fuel being moved the short distances to the aircraft via vacuum-jacketed pipeline with cold gas boil-off return lines. The gas used could be generated or produced wherever it was most convenient and then piped in its gaseous form to the airport-associated liquefier. Considering the global nature of our air transportation system, worldwide conversion to a cryogenic alternate fuel would be required.

In the case of the automotive and ground transportation, conversion to cryogenic fuel could be somewhat more localized. The development of small efficient liquefiers for connection to the gas pipeline delivery system is a possibility. Large central liquefaction plants with delivery to the vehicle filling station by rail or trailer are also conceivable methods. While either of these methods could be utilized, it should be recognized that my estimation of the liquid hydrogen production capability necessary to replace the existing gasoline production for automobile use in the U.S. would require approximately 100 producing plants, each with a capacity of 3000 tons per day. This is compared to a total U.S. capacity today of approximately 150 tons per day.

As I see it, the problems associated with the storage and distribution of cryogenic fuels have been solved within the capability and the technology of today's designers of this type of equipment. As improvements are made in the technology and economics of production and utilization, the storage and distribution equipment designer will be called upon to perform in his current area of expertise, and will be expected to achieve the performance improvements that will result from increasing utilization of his art.

Comments on Timmcke's Paper

The main problem with flatbottom storage for liquid hydrogen is the heat transfer through the bottom load-bearing part of the system. A vacuum-powder insulation system cannot be designed as a foundation, and what could be designed would have fairly good heat-transfer properties. However, it is possible, since hydrogen is so light a liquid, to build a very large spherical or vertical cylindrical container up to 150 feet in diameter, and vacuum-insulate it.

•

The current electrical energy requirement for liquefying hydrogen is about 7 kilowatt hours per pound of hydrogen—possibly slightly, but not appreciably, lower as the plants get up in size. The current price of liquid hydrogen (LH_2) is about $1.44 per pound. As for the energy equation, it takes about 50,000 Btu to liquefy a pound of hydrogen, and with electricity being generated at 10,000 Btu per kilowatt, it works out just about even: it would take about a pound of hydrogen to liquefy a pound of hydrogen. Therefore, in order to show a reasonable energy balance when using LH_2 as a fuel, one must also make use of its low-temperature properties. In other words, since it has both heat of combustion and heat-sink properties, the systems engineer can make double use of the fuel—once as a heat producer and once as a heat absorber. This is one of the reasons that LH_2 is so attractive as an aircraft fuel. Its heat absorption properties can be used to get better engine efficiencies. Automotive engineers, however, have not yet begun thinking along those lines.

Energy is equally valuable in the form of a potential low-energy sink as in the form of a potential high-energy source. For example, the liquid natural gas (LNG) that would be landed in Maryland and in Georgia holds about as much refrigeration capability as the entire industrial gas processing industry: oxygen, hydrogen, argon, helium, and the rest.

B. Storage and Distribution of Synthetic Fuels

Jerry E. Berger
Shell Oil Company
Houston, Texas

This discussion will deal with storage and distribution of non-cryogenic, "conventional" synthetic fuels. At the outset, it will be useful to limit the scope of this paper by confining it to reasonably foreseeable alternate synthetic fuels for use in the transportation industry. In particular, the focus will be on synthetic fuels for use in automotive applications; the time frame of interest here is on the order of fifteen years or so.

Table 15 contains a list of selected "synthetic fuels" which, in the view of the author, represent reasonably foreseeable alternative fuels for the near-term future. In addition to the alcohols, some others will be discussed and, finally, attention is directed to synthetic hydrocarbons derived from coal, from shale, or from synthesis gas

Table 15   Selected "Synthetic Fuels" for Transportation Purposes

	Energy Content per Unit Volume (Gasoline = 1)	Solubility in Water	Boiling Point (°F)	Flammability Limits (vol. %)
Methanol	0.5	∞	149	6.7-36
Ethanol	0.7	∞	172	4.3-19
Higher Alcohols (C_3,C_4)	...	∞	>180	~2-10
Methyl-*t*-Butyl Ether	0.8	Slight	131	...
Diisopropyl Ether	...	Slight	156	...
Ethyl-*t*-Butyl Ether	...	Slight	158	...
Hydrocarbons				
From Coal Liquefaction	~1	Nil	Range	Approximately
From Shale Retorts	~1	Nil	Range	1.4-7.6
From Fischer-Tropsch Synthesis	~1	Nil	Range	(typical
From Methanol via Mobil's				for
Process	~1	Nil	Range	gasoline)

obtained from a variety of possible sources. Selected properties which are relevant to subsequent discussion also are included in Table 15.

Since the primary concern here is automotive fuel, it is of interest to look at the distribution of gasoline consumption across the United States and compare those figures with the location of petroleum refining capacity in the United States. Table 16 contains some relevant data. In Table 16, it can be seen that three states, Texas, California, and Louisiana, combine to provide more than half the gasoline which is produced domestically. These same three states, however, account for less than a fifth of the gasoline consumed in the United States. Pennsylvania and New Jersey are about in balance with regard to gasoline production and consumption, but most other states consume far more than they produce. Some states have virtually no refining capacity. Hence it is obvious that gasoline production does not coincide with gasoline consumption, and, therefore, transportation and distribution are essential to move the fuel from the refinery where it is produced to the centers of gasoline consumption.

Figure 17 is a schematic diagram which illustrates the ways in which transportation fuels move from a refinery through the distribution system to the ultimate user. When transportation fuel leaves the refinery, the principal routes of movement are by pipeline, by ship, and by barge. These various transportation routes converge on strategically located terminals where product is held until needed. From terminals, transportation fuel moves, for the most part, by tank trucks which supply service stations or fleet accounts. Figure 18 is a particularized drawing which shows the distribution network for one refiner who has refineries located on the Gulf coast and in the Midwest near the Mississippi River. The importance of water transport is the key feature which Figure 18 illustrates. For example, the two Gulf coast refineries provide transportation fuel for outlets on the Gulf coast as far as the tip of Florida; these terminals are served by barge. In part, these barges move through the so-called Inland Waterway

Table 16   Gasoline Production and Consumption

	% of U.S. Refining Capacity [a]	% of U.S. Gasoline Consumption [b]
Texas	27.3	7.3
California	14.1	9.8
Louisiana	12.4	1.8
Illinois	7.0	4.9
Pennsylvania	4.8	4.4
New Jersey	3.1	3.1
44 Other States	31.3	68.7

[a] As of Jan. 1, 1978. Source: *Oil and Gas Journal*, March 20, 1978, p. 113.
[b] For 1977. Source: *1977 National Petroleum News Factbook*.

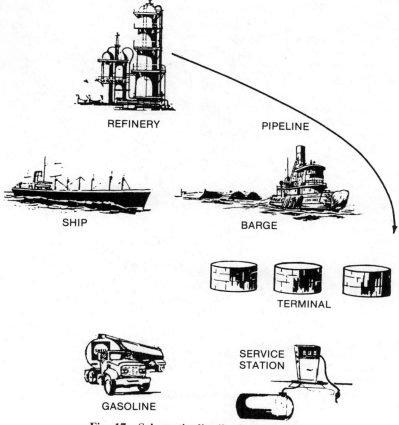

Fig. 17 Schematic distribution system

stretching along the Gulf coast as far as the Panhandle of Florida. The particular oil company whose operations are shown on Figure 18 has no East coast refinery. Hence, in order to supply East coast markets, fuel from the two Gulf coast refineries is carried by ocean-going ships which round the tip of Florida and serve terminals up and down the Atlantic seaboard by discharging cargo at major ports. These ports stretch from Georgia to Massachusetts. One of the Gulf coast refineries and the midcontinent refinery are located on the Mississippi River. Product can move by barge up and down the Mississippi and Ohio River systems from these two refineries. In addition, product pipelines move fuel from the refineries to centers of population.

Almost invariably, ship or barge transportation involves a wet system in contrast to an anhydrous one. The degree of water contamination accompanying marine shipments is highly variable. Some

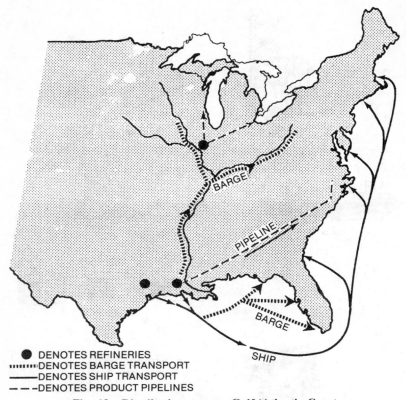

● DENOTES REFINERIES
▪▪▪▪▪▪▪DENOTES BARGE TRANSPORT
———DENOTES SHIP TRANSPORT
— — —DENOTES PRODUCT PIPELINES

Fig. 18   Distribution system: Gulf/Atlantic Coast

shipments of product arrive at their destination in an essentially
anhydrous state, but others are contaminated with relatively large
quantities of water. In some instances, as much as several inches of
water can be introduced to a storage tank during receipt of the
shipment. There are several sources for this water which accompanies
marine shipments, but historically the principal source has to do with
ballast. Empty ships and barges frequently use water as ballast for
their return or empty leg of the voyage. This ballast may not be
completely removed prior to loading product for a subsequent voyage
and hence a substantial quantity of ballast water may accompany the
succeeding cargo.

Incomplete removal of ballast water is not the only source of water
contamination in the transportation fuel distribution system. Other
sources include rainwater which may enter through the seals of
floating-roof storage tanks or may enter the underground service
station storage vessels if a tank seal is improperly fitted following

receipt of product. It is generally recognized that water is present in varying amounts in today's gasoline distribution system. Few quantitative data have been published, however, on this subject. One recent survey looked at 300 service stations and about 100 bulk tanks located in various climates in the western United States. For service stations, the median value for half-filled tanks was a tenth of a percent of water. For bulk tanks, the median value was nine-tenths of a percent water. Most water in the overland distribution system is thought to enter from rain passing through the seals of floating roof tanks. It has been estimated that about 4% of the rain falling on the roof of a floating tank enters the product contained in the tank. [23]

At the refinery, water can be carried over from processing steps. Some hydrocarbon streams are quite warm when they leave a process unit and, if these streams are subjected to aqueous wash procedures, the fuel which emerges is water-saturated at a relatively high temperature. When this product enters the distribution network, it cools and the water which precipitates must be removed at the terminals. In addition, diurnal breathing of storage vessels, especially in humid environments, can introduce moisture into the system which can condense and collect in the bottom of the vessel.

Despite the fact that the fuel transportation system today is a wet system, problems are rarely encountered. This is because gasoline or diesel fuels are capable of dissolving relatively little water and any water contamination which exists simply falls to the bottom of the storage vessel. From here it can be removed periodically.

With this brief description of the transportation fuel storage and distribution system, it may now be appropriate to return to the "synthetic fuels" which are listed in Table 15. It is instructive to consider each of the candidate synthetic fuels contained in Table 15 in the context of the characteristics of the distribution and storage system which exists. Methanol and ethanol are frequently mentioned as potential gasoline blending components but blends of methanol or ethanol in gasoline would not be compatible with the distribution system which exists today. Figures 19 and 20 present phase diagrams for the methanol-gasoline-water and ethanol-gasoline-water systems. It is apparent that blends of gasoline and either alcohol are relatively intolerant of water contamination. This is significant because the distribution system from the refinery to consumer which exists today is a wet system and, if an attempt were made to introduce alcohol/gasoline blends at the refinery and transport them through the distribution system to the customer, water contamination would cause the blends to separate into two phases.

Figure 21 illustrates the phase separation temperatures for a variety of methanol/gasoline blends. The parameter here is the percent of

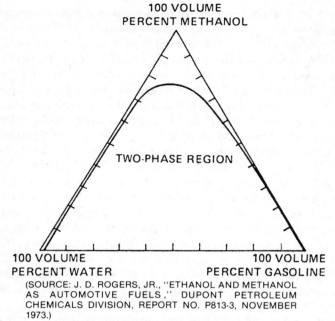

(SOURCE: J. D. ROGERS, JR., "ETHANOL AND METHANOL
AS AUTOMOTIVE FUELS," DUPONT PETROLEUM
CHEMICALS DIVISION, REPORT NO. P813-3, NOVEMBER
1973.)

Fig. 19 Phase diagram for the system methanol-water-gasoline at 77°F

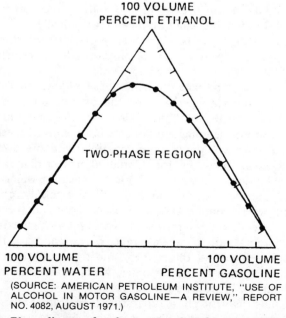

(SOURCE: AMERICAN PETROLEUM INSTITUTE, "USE OF
ALCOHOL IN MOTOR GASOLINE—A REVIEW," REPORT
NO. 4082, AUGUST 1971.)

Fig. 20 Phase diagram for the system ethanol-water-gasoline at 76°F

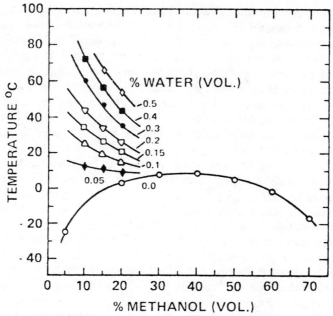

Fig. 21 Phase separation temperature for a methanol/gasoline blend in the presence of water

water present in the blend. As greater amounts of water are added to a methanol/gasoline blend, higher temperatures must be maintained in order to keep the system in solution.[24] In contrast, water contamination rarely causes problems with hydrocarbon fuels. Water is virtually insoluble in such fuels and it settles innocuously to the bottom of storage tanks. Fuel is withdrawn through lines positioned above the level of the fuel/water interface and provision is made for periodic withdrawals of accumulating water as necessary.

In theory, the water incompatibility problem could be mitigated, in part, by transporting the gasoline and methanol or ethanol to the service stations separately. Following blending at the pump, a relatively anhydrous blend might be introduced to a vehicle tank. The disadvantages of this approach have to do with the necessity for building segregated facilities for the alcohol component of the blend and for guarding against water contamination of the alcohol during its transport and storage.

In theory, alcohols such as methanol or ethanol could be used in the "pure" state without blending into gasoline. Such fuel might be acceptable in dedicated fleets: for example, dedicated fleets of taxi cabs in major metropolitan areas. In such circumstances, the cost of segregated facilities would be minimized when compared to general nationwide distribution of alcohol-gasoline blends. Because a given volume of methanol or ethanol has a substantially lower energy content than the same volume of gasoline, it would be necessary to enlarge the storage facilities to compensate for this difference. A potential problem would be the occasional unscrupulous operator who would deliberately dilute his methanol or ethanol with water.

The water tolerance of alcohol-gasoline blends is highly dependent upon the molecular weight of the alcohol. Water sensitivity is greatest in the case of methanol-gasoline blends and decreases as higher alcohols are used. Indeed, at least three oil companies now are using several percent of tertiary butyl alcohol as a gasoline blending component. This strategy is motivated in part by octane-quality considerations and in part by the availability of by-product tertiary butyl alcohol. Apparently the addition of tertiary butyl alcohol is being accomplished without undue difficulty.

Certain low molecular weight ethers also represent potential gasoline blending components. Included in this category are methyl tertiary butyl ether, diisopropyl ether, and ethyl tertiary butyl ether. Inclusion of these materials in this discussion is warranted because the petrochemical industry in the future is likely to produce relatively large volumes of isobutylene and propylene as by-products from the crackers designed to produce ethylene. One possible way to dispose of a surplus of isobutylene or propylene is to convert these olefins to ethers using low molecular weight alcohols such as methanol or ethanol. If methanol were available in sufficient supply, for example, it would be possible to react methanol and isobutylene to form methyl tertiary butyl ether. A plant having a capacity of 100,000 tons per year of methyl tertiary butyl ether has been in existence in Italy since 1973.[25] When compared to blends of alcohols and gasoline, blends of low molecular weight ethers and gasoline have several advantages. The low molecular weight ethers are miscible in all proportions in gasoline. In addition, such ethers are only slightly soluble in water, and blends of low molecular weight ethers in gasoline show no phase separation problem such as those existing with methanol-gasoline and ethanol-gasoline blends. Although diethyl ether is known to form explosive peroxides readily, available information indicates that peroxides are not formed with methyl tertiary butyl ether. With regard to diisopropyl ether and ethyl tertiary butyl ether, it is reasonable to speculate that peroxide formation probably occurs under some circumstances. The blending properties of low molecular weight ethers such as methyl tertiary butyl ether are similar in most respects to the relatively low molecular weight gasoline components which are used now. Hence, distribution and transportation problems with these materials blended into gasoline are likely to be minimal.

In addition to the oxygenated species which have been discussed earlier, there exists an array of synthetic hydrocarbons which could represent candidate transportation fuel components. In theory, coal liquefaction produces a petroleum-like material which can be refined in conventional equipment. The result of such processing is a gasoline which is similar in many respects to today's gasoline. Alternatively, retorting shale produces a second petroleum-like liquid which also is amenable to further refining in existing equipment. Performing this sequence of operations also produces a gasoline which resembles today's product.

Somewhat more remote possibilities include the use of the Fischer-Tropsch process for the conversion of synthesis gas to hydrocarbons. In theory, the synthesis gas could come from a variety of sources including shale, coal, or organic matter of a renewable type. A final alternative involves the synthesis of methanol from synthesis gas and the subsequent conversion of the methanol to a gasoline-like hydrocarbon blend using Mobil's zeolitic catalysis approach. This process is said to yield a mixture of high-octane hydrocarbons in the appropriate boiling range and to resemble good-quality gasoline blending stocks.

All these synthetic hydrocarbons, regardless of their source, warrant little further attention in the discussion. The final product of coal liquefaction or shale retorting or Fischer-Tropsch synthesis or the Mobil process is a material which resembles today's gasoline components in many respects. Hence, the inclusion of such materials as a gasoline blending stock is likely to present few, if any, problems. Indeed, it should be recalled that much of today's gasoline barrel consists of synthetic fuels since the components have been chemically altered to a substantial degree. For example, about a third of today's gasoline barrel is catalytically cracked gasoline. Perhaps another third of today's gasoline barrel consists of aromatic species which were originally alkanes prior to their cyclization and dehydrogenation. Additional synthetic gasoline components include alkylate and isomerized paraffins. The synthetic hydrocarbons listed in Table 15 are likely to present no more problems as gasoline components than the existing synthetic components found in today's gasoline blends.

The previous discussion has concentrated on handling problems which arise when some synthetic fuel components are added to gasoline and the blend is transported through today's distribution system. There are other factors to be considered, however. One of these is a potential environmental problem and stems from a rather simple operation in today's practice. As water accumulates in the gasoline distribution system, periodically it is withdrawn and disposed of in an environmentally acceptable manner. In the case of water-soluble gasoline-blending components such as methanol or ethanol, the practice of discarding water bottoms becomes considerably more complicated. If phase separation has occurred in a storage vessel, it is likely that the bottom phase will consist of aqueous alcohol and discarding this solution has potentially harmful environmental impacts. It would be necessary to collect such aqueous alcohol solutions and subject them to further treatment in order to render them innocuous.

Another problem has to do with corrosion. If alcohol-gasoline blends are contaminated with water and if phase separation occurs, the resulting aqueous alcohol phase is relatively corrosive to some metals such as magnesium and aluminum. Hence, it would be necessary to change those components of the distribution system which are fabricated of susceptible metals.

A detailed discussion of the toxicological properties of blends of synthetic fuels in gasoline is beyond the scope of this paper. However, there is one area that deserves mention, and that area is the possibility of spills. Gasoline spills or diesel fuel spills represent serious occurrences; however, a spill of a methanol-gasoline blend would represent serious consequences of a different type. This is because, if the spill entered a waterway, the methanol would readily dissolve in the water and could conceivably enter a drinking water supply. In any event, fish and wildlife would be jeopardized.

In summary, the use of "synthetic fuels" for transportation purposes may or may not present special problems to the transportation and distribution network which now exists. Those synthetic hydrocarbons which resemble today's gasoline components would offer few additional problems. Some oxygenated species such as the ethers appear to be free of major problems, but other oxygenated species such as methanol or ethanol present special problems to those persons involved in the transportation and distribution of such fuel blends. Transportation and distribution of methanol or ethanol as the pure component might avoid some of the problems associated with gasoline-alcohol blends, but this strategy has the disadvantage of requiring segregated facilities.

Alcohols or alcohol-gasoline blends could be injected into the nationwide fuel supply network, but doing so would impose major logistical costs and would require extensive new investments in the blending and storage and transportation facilities. The question then becomes one of balancing the costs against any net societal gains which such new fuels might provide.

Comments on Berger's Paper

One of the things that ought to be said about the gasoline distribution network as it now exists is that it is an aging network.

Many tanks at service stations have been underground for twenty years or more and in many cases are no longer owned by the people who had them installed. Their condition and integrity are often in question. Leaks caused by galvanic corrosion and ground shifting, while not catastrophic, are sometimes serious. While this subject does not come under the purview of the Symposium, the fact should be recognized in any discussion of alternative fuels which may be utilized in blends with existing components.

C. Summaries of Prepared Remarks and Commentary

1. James L. Keller

Methanol, along with hydrocarbons from oil shale and coal, is among the most promising potential alternative fuels; we should be learning how to live with it should we have to. Most discussions of methanol these days center on methanol-gasoline blends, probably because this seems a simple and appealing way to use methanol. But it is not necessarily the best way. Some of the problems already mentioned with blends include phase separation and corrosion. There are others:

(a)   They increase the vapor-lock tendency of gasoline. To control it one must pull butane out, thereby adding less fuel than was probably intended.

(b)   They have a negative effect on car drivability, unless carburetors are readjusted for the blend.

(c)   They tend to "lean out" the air-fuel ratio of the car, with possibly negative effects on emissions. (A recent EPA ruling requires a waiver under the 1977 Clean Air Act amendments for alcohol use in automobiles.)

(d)   In blends, alcohol tends to loosen dirt and sediments in the distribution system and carry them along to plug up filters, etc.

(e)   Blends have undesirable effects on various materials used in fuel systems, especially plastics and synthetics; for example, polyurethane disintegrates in the presence of methanol-gasoline.

To get even partial control of the phase separation problem mentioned earlier, one must add considerable amounts of higher alcohols as cosolvents for the methanol. These would either be bought at high price or made along with the methanol, which would require the development of new technology.

The benefits of the blends are increases in octane rating—the blending motor octane value is about 110—and an increase in fuel volume. But for an increase in volume of 10 to 20% methanol—which is the best one can do within drivability limits—one adds only 5 to 10% energy. And after backing out butane to offset the vapor-lock problem, one ends up with an energy increase of only 2 to 4%. Put another way, the additional energy is only about 20% of the added methanol.

With so many disadvantages, and such small gains, it is not surprising that independent studies conducted by Professor Jay Bolt and his coworkers from the University of Michigan, by General Motors, and by Union Oil of California have come to about the same view: that methanol is best used straight, in distribution systems and in cars that are designed for it. The advantage here is that ultimately, as our gasoline supply dwindles, we would have a complete replacement fuel.

Methanol used straight is an excellent fuel. Its very high octane rating would permit compression ratios up to 13 or 14 and the development of concomitantly high engine efficiencies. And of course there would be no problems of phase separation. Some of the problems that *are* associated with methanol can be controlled by spiking it with gasoline or light hydrocarbons: cold starting, explosive vapors in the tank, and its practically invisible flame as it burns. There are other problems in the distribution system that must be worked out; e.g., a need for alcohol fire-fighting foams, and a greater incompatibility for straight methanol (rather than blends) with some materials in the system. This might be done as a result of experience with small captive fleets—taxicabs, post-office fleets, or the like, as suggested by Berger—that are either designed or retrofitted to handle straight methanol. [It should be pointed out that separate distribution systems are not new to the petroleum industry. It now separately handles leaded and unleaded gasolines, liquefied petroleum gas (LPG), diesel fuels, jet fuels, and black residual fuels.]

But ultimately, should we go to large-scale use of methanol, we must expect approximately double the investment in storage tanks, pump stations, and pipe lines to handle a fuel with half the energy content of gasoline. We would also have to replace the fiberglass tanks used in so many service stations. Such a changeover would be a major undertaking for the petroleum industry. However, the lack of an alternative fuel would be more than a major undertaking. It would be a disaster—not just for the petroleum industry, but for the entire economy.

Ethanol, which is being promoted mostly by those who see it as a new market for farm products, is, economically, distinctly less attractive than methanol. The amount of source material, at least as compared to coal, is much more limited. As for performance, ethanol blends seem to suffer from much the same problems as methanol blends, but to a much lesser extent: phase separation, drivability loss, air-fuel effects, vapor-lock increase, etc. It is, however, ecologically appealing: it can be made from renewable resources and thus could, in some circumstances, add to the energy supply without drawing on fossil reserves and without adding CO_2 to the atmosphere. The gasoline-ethanol blend called Gasohol is now being sold in a few service stations; we might call this a kind of marketing experiment, and we will watch to see how the claims made for Gasohol hold up against real-life economics and actual performance for real customers.

2.   William J. Koehl

The technical and economic problems associated with manufacturing and using synthetic fuels have been discussed thoroughly in other sections of this volume. The interface between manufacture and use—namely storage and distribution—is also fraught with technical and economic problems. Berger, of Shell Oil Company, has reviewed the changes in storage and distribution practices, from those used today for petroleum fuels, that would be required for alcohols and for blends of alcohols in gasoline. To recapitulate, the major problems include:

(a) handling twice as much methanol as gasoline to supply equal energy,

(b) modifying distribution systems to avoid contamination by water which causes alcohols to separate from blends,

(c) replacing methanol-sensitive components in fuel distribution systems and in existing vehicles, and

(d) rejecting butane from the gasoline pool because of methanol's high vapor-forming tendency in the blend.

All of these requirements add to the cost of supplying alcohol fuels. However, there is a way to circumvent these handling requirements, and to do so at less cost, by converting the alcohol to a gasoline that is completely compatible with present gasolines.

Mobil Research and Development Corporation is developing a process for converting methanol into high-octane gasoline (see Refs. 26 and 27 for background). This work is being done under cost-sharing contracts with the U.S. Department of Energy. The process uses a zeolite catalyst of the ZSM-5 class that essentially dehydrates methanol, giving high-octane gasoline and water as the primary products. The Mobil process could be integrated into a plant producing methanol from coal (or any other resource) as illustrated in Figure 22, while adding less than 10% to the investment cost for the process sequence.

The Mobil process has been operated in a fluid-bed pilot unit with a feed capacity of 4 barrels per day. Typical product yields are given in Table 17. On a hydrocarbon basis, the raw C_{5+} gasoline fraction amounted to 60 wt.%. Alkylation of propylene and butene with isobutane from the process increased the gasoline to 88 wt.% of the

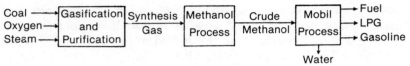

Fig. 22   Gasoline from coal

hydrocarbons. On an energy basis, the hydrocarbon products account for approximately 95% of the thermal energy in the methanol feed.

The gasoline produced is of very high quality, as shown in Table 18. Its octane number exceeds current requirements for unleaded gasoline. Its composition and boiling range are similar to conventional gasoline, and it contains no sulfur or nitrogen, since there is none in the feed.

We estimate that the cost of converting methanol to gasoline will be substantially less than the cost of displacing butane from gasoline, replacing sensitive materials in cars, and changing storage and distribution systems in order to supply blends of methanol in gasoline. In addition, conversion would assure the same reliable performance as conventional gasoline in distribution systems and in vehicles, and would avoid the risks of unforeseen problems in diverse vehicles or of accidental contamination with water. It would also obviate the need to educate fuel handlers and users about the different safety, health, and product quality requirements of methanol fuels. Thus, we believe

Table 17   Yields from Methanol in Mobil 4 B/D Fluid-Bed Pilot Unit

Yields, wt. % of Methanol Charged	
Methanol + Ether	0.2
Hydrocarbons	43.5
Water	56.0
CO, CO_2, Coke	0.3
	100.00
Raw Hydrocarbon Products, wt. %	
Light Gas	5.6
Propane	5.9
Propylene	5.0
Isobutane	14.5
n-Butane	1.7
Butenes	7.3
C_5+ Gasoline	60.0
	100.0
Finished Fuel Products	
Gasoline	88.0
LPG	6.4
Fuel Gas	5.6
	100.0

Table 18   Typical Properties of Gasoline from Methanol

Composition, wt. %		
Paraffins	56	
Olefins	7	
Naphthenes	4	
Aromatics	33	
	100	
Octane		
Clear	96.8 (research)	87.4 (motor)
Leaded (3 cm³ tel/gal)	102.6 (research)	95.8 (motor)
ASTM Distillation, °F		
10%	116	
50%	217	
90%	336	
REID Vapor Pressure, psi	11	
Specific Gravity	0.730	
Sulfur	Nil	
Nitrogen	Nil	

conversion of methanol to gasoline is preferable to blending it in gasoline.

3.   Symposium Participants

(a) EPA's 1977 "clean air" amendments require that no fuels or fuel additives be put on the market if they are not substantially similar to those used in the qualification tests for 1975 and later automobiles. EPA has not granted a waiver from this for Gasohol, declaring that, since it was not substantially similar, it would have to prove its qualifications. EPA *has* granted a temporary waiver on the drivability tests for Gasohol, but as for the rest, Gasohol changes the air-fuel ratio in a carburetor designed to operate on gasoline, thus affecting the emission control as well as upsetting some of the fuel-system components.

(b) The factor that for now prevents the Mobil conversion process from being the answer is cost. The cost of converting 2.4 gallons of methanol into 1 gallon of gasoline would add about 10 cents to the price of the gallon of gasoline. However, this figure may become competitive when weighed against the additional costs involved in running automobiles on gasoline-methanol blends: the costs of changing the fuel distribution system and the fuel systems in the cars themselves. The added 10 cents per gallon on gasoline from methanol is pretty much on a line with costs of distribution if we were to go with a 10% methanol blend; i.e., it is about 1 cent additional per gallon of blend compared with 1 cent additional per gallon of gasoline, when one tenth of that gallon is gasoline converted from methanol.

(c) It is difficult to make a direct comparison between the current systems of transporting petroleum—that is, through the vast network described by Berger— and transporting ethanol, which would be from a multiplicity of local agricultural sources to closer destinations. However, if the primary alternative is methanol, then very large conversion plants would probably be built close to the coal mines, and from the plants the methanol would travel by pipeline. This system could be even more centralized than the petroleum industry's.

(d) Investigative work on ethanol as an alternative fuel will proceed along similar lines as that done on methanol, but the scope will be smaller: the work will be primarily with blends. With ethanol there is the additional problem of combating or preventing the extraction and bootlegging of the ethanol from the blend.

(e) *Question:* Brazil appears to have no difficulty distributing blends of up to 20% ethanol in gasoline. Why can't U.S. experience be expected to be similar.

Answer: For several reasons. First, average ambient temperatures in Brazil are higher than in the U.S.; consequently, the water tolerance of blends is greater. Second, the gasoline distribution system in Brazil affords fewer opportunities for exposure to water because it is largely a road-transport system; while the U.S. system is a complex network combining pipeline, marine, rail, and road segments. Furthermore, to minimize the risk of exposure to water, Brazil blends ethanol into gasoline at terminals immediately prior to delivery to retail outlets.

V.
ROAD VEHICLE UTILIZATION

A. Outlook for Performance of Alternative Fuels in Automobiles

Serge Gratch
Ford Motor Company—Research
Dearborn, Michigan

1. Abstract

This paper reviews fuel requirements of conventional and advanced automotive engines and attempts to predict how future fuels will match these requirements. It concludes that Otto and diesel engines will remain dominant. They will be best suited for liquid hydrocarbon fuels, with little or no change in volatility, stability, octane/cetane, and other requirements. However, they could be adapted to alternative fuels. Direct injection, stratified charge engines will have less stringent octane and volatility requirements. Stirling and gas turbine engines will be extremely flexible in this respect.

2. Introduction

On the basis of the literature and of the papers previously presented at this Symposium, it seems clear that petroleum will not continue to be the main source of transportation fuel indefinitely. The main question at present seem to be the rate of introduction of alternative fuels and their selection. Although the objective of this paper is to examine the factors affecting this selection from the point of view of the automobile, it is impossible to address this issue rationally without recognizing some important limitations imposed by the fuel distribution and storage systems. Specifically, any alternative fuel which is incompatible from the point of view of distribution and storage with the automotive fuels currently in use would face a very serious handicap. During the transition period from present fuels to the new fuel, a new dual storage and distribution system would be required. The time required for transition would likely be very long since the average life of automobiles is approximately 10 years and, on the basis of previous studies, it seems unlikely that the cars already in use could be retrofitted economically to be adapted to the new fuels. Therefore, in this paper those fuels which would require a new distribution system will be discussed only briefly since their adoption appears unlikely in the face of such an added economic burden.

For the purpose of this paper it is also useful to recognize that the spark ignited engine will remain the main prime mover for passenger cars for a long, long time. No other engine combines so many advantages: low cost, reasonable torque-speed characteristics, moderate emissions, suitability for convenient fuels, and, above all, great versatility. Some of these characteristics have been unfairly maligned. For instance, it has been claimed that the falling torque with decreasing speed below some critical value is a disadvantage, since it requires a costly transmission, which would not be needed with engines having the torque-speed characteristics of Rankine cycle steam engines or series wound electric motors. The point that is often forgotten is that the torque-speed characteristics of the Otto cycle engine provide a very important safety feature, since, in the case of many types of accidents, the engine tends to stall, thus decreasing the potential for further damage. Similarly, the octane requirements of Otto cycle engines have been considered undesirable, and rightly so. But is the cetane requirement of diesel engines not worse? Several refining processes capable of increasing octane exist, while it is very difficult to increase cetane.

The versatility of the Otto cycle engine, which allows tradeoffs among fuel economy, performance, emissions, manufacturing cost, etc., is also a great advantage of this engine. This engine operates satisfactorily over a wide range of temperature, loads, and speeds. Its versatility has been amply demonstrated by the impressive reductions in vehicle emissions and, more recently, by the gains in fuel economy, which have been achieved by relatively minor, although oftentimes costly, modifications of that engine. Because of this versatility it appears likely that this engine will survive and that, accordingly, the fuel requirements of the future will be determined mainly by the needs of the Otto cycle engine.

Aside from these advantages of the spark ignited engine, it must be recognized that the transition to a drastically different engine would be a slow process. A few years of lead time would be required even for the initial introduction of any drastically different engine. Presumably the initial introduction would consist only of one engine line, with a maximum capacity probably of the order of half a million units a year. Additional engine lines would be produced only gradually since the capacity of the manufacturers of transfer lines and other machinery required for the production of engines is limited and a complete retooling of engine manufacturing probably would require between 10 and 15 years. Thus, complete conversion to the production of drastically different engines is not likely to be completed before the year 2000. Another 10 years would elapse before the previously built automobiles with conventional engines were substantially eliminated from the total car park. Thus it appears likely that fuels suitable for conventional spark ignited engines will be required at least until the year 2010.

Another consideration in the following discussion is that any alternative fuels and alternative engines introduced in the future must allow compliance with the Clean Air Act and with the Energy Policy and Conservation Act. These two laws mandate very stringent restrictions in vehicle emissions and requirements for fuel economy.

With these considerations as a background, the paper will review briefly the outlook for various alternative fuels relative to the different engines currently of interest, starting with the conventional spark ignited engine and continuing with modified spark ignited engines, diesel engines, Stirling, and gas turbine engines.

3. Conventional Spark Ignited (Otto) Engines

Otto cycle engines have been operated with a wide variety of fuels, such as hydrogen, ammonia, methane, methanol, ethanol, and various blends. Of these, the gaseous fuels are unattractive for general use since they would require replacement of the huge gasoline distribution system in current use. In addition, hydrogen is unattractive at present because of the poor energy efficiency of the established processes for its manufacture and because its storage entails major weight penalties. Ammonia is undesirable because of its toxicity and irritant properties; moreover, very close control would be required to avoid the emission of objectionably high amounts of oxides of nitrogen. Methane is unattractive because of the current shortage of natural gas; were it to be produced from coal or other abundant resources, it would seem preferable to synthesize in its place a liquid fuel, in order to avoid the need for a new fuel distribution system. Propane cannot be properly described as an alternative fuel since in the form of LPG it has been in limited use for quite a long time. Its use is not likely to increase because of its limited availability.

Nevertheless, some additional comments on hydrogen and methane are appropriate. Hydrogen has been studied[28] extensively as a fuel for the spark ignited internal combustion engine. Although modifications are required, these appear to be relatively straightforward. Its high flame speed allows hydrogen to burn over an extremely wide range of fuel air ratios.[29] This in turn allows operation at a much leaner fuel air ratio. The combination of the high flame speed and of the leaner operation results in significant improvement in thermal efficiency.[30] On the other hand, these combustion characteristics require modification to prevent combustion harshness and flame flashback into the intake manifold.[31] As indicated above, from the point of view of usage in an automobile the main disadvantage of hydrogen is the bulkiness required for its storage. [32] Three means of storage may be considered: as a compressed gas, as a cryogenic liquid, or in the form of a metal hydride. In any of these forms the total weight and volume of the fuel with its container would be much larger in the case of hydrogen than in the case of gasoline. In terms of volume it is estimated that the penalty of hydrogen would be at least a factor of four.

Methane is more similar in properties to conventional gasoline. Although gaseous at room temperature, it can be liquified much more easily than hydrogen since it is a liquid at atmospheric pressure at a temperature of $-162\&C$. Its density as a liquid is almost as high as that if gasoline so that, as a result, including the added bulk of the cryogenic container, its volume requirements would be expected to be only about double those of gasoline. In addition to the change in storage systems, the main modification required to a vehicle for the use of methane would be the addition of means for vaporization and for the use of a gaseous fuel in the carburetor. Because of the low temperature of storage, it is probably impractical to feed liquid methane directly into the carburetor. Methane has a relatively high octane number and is reported to burn somewhat more cleanly than gasoline. Unfortunately, it may suffer an emission penalty with advanced emission control systems because of a peculiarity in present vehicle emission regulations. Methane has been found to be very unreactive on the oxidation catalysts currently in use in most automobiles. [33] This is not of real concern in terms of air quality since methane is relatively harmless and it is not reactive in photochemical smog formation. Unfortunately, however, current vehicle emission standards for hydrocarbons are in terms of total hydrocarbons including methane. Thus it may be difficult to meet future hydrocarbon vehicle emissions standards with a methane fueled engine.

Renewed interest has been shown in recent years in alcohols, particularly methanol and ethanol. An extensive review of methanol as an automotive fuel has been published recently. [34] Studies show a mixture of advantages and disadvantages for this fuel, [35, 36] as well as for ethanol. [37] Both methanol and ethanol have excellent octane quality. They have been reported to burn with greater efficiency and to produce more power at an equal equivalence ratio than gasoline. However, alcohols have a lower heat of combustion than gasoline, as shown in Table 19.

As a result, a vehicle fueled with methanol requires roughly twice the volume of fuel to drive an equivalent distance. One fueled with ethanol requires one and a half times as much fuel as gasoline for the same distance.

The reason that, in spite of the lower heat of combustion, alcohols develop more power is a combination of two factors. First of all,

Table 19   Heat of Combustion of Various Fuels

Fuel	Btu/lb	Btu/gal
Methanol	8,570	56,560
Ethanol	11,500	75,670
Gasoline	18,900	115,400

alcohols require less air for combustion, in approximately the same ratio as their heat of combustion. Moreover, alcohols have a much larger heat of vaporization than gasoline. This results in a considerable cooling of the intake mixture, which in turn results in a higher density of the mixture and correspondingly in a higher total charge. Therefore, the energy available for each stroke is greater than in the case of gasoline.

The intake mixture cooling in combination with the higher flame speed of alcohols is probably responsible for their higher octane number. This cooling, however, combined with volatility differences, is probably responsible for greater difficulty in cold starting and for poorer cold drivability with these fuels. Other problems with alcohols and alcohol-gasoline blends include a higher rate of corrosion of some materials, due in part to the greater affinity of alcohols for water. Moreover, alcohols cause softening and swelling of certain plastics. In the case of blends, an additional problem is the tendency to separate into two phases at low temperature when water is present as a contaminant. Although none of the disadvantages are insurmountable, general acceptance of these fuels is unlikely until more economical means are developed for their production.[37-39]

Previous writers have discussed in detail the prospects for fuels from coal, shale, etc. It suffices here to say that apparently it will be possible to obtain from these sources fuels which will be quite similar to conventional gasoline.[40] Possibly it will be more efficient to produce somewhat different fuels, such as methanol, from these sources, but this has still to be determined. It may be noted that different hydrocarbons lead to the production of different amounts of exhaust pollutants.[41] This effect appears too small to be of practical concern for petroleum-derived gasolines, but must be kept in mind in the case of alternate fuels.

4.   Modified Otto Cycle Engines

Although many modifications of the Otto cycle engine are possible and deserve consideration, the most interesting class apears to be the direct injected, stratified charge (DISC) engine, as exemplified, for instance, by the Ford PROCO and the Texaco TCP.[42-44] The great interest in these engines has been generated by their potential for significant improvements in fuel economy. They also have the potential of much greater tolerance for variations in fuel properties than is possible with conventional Otto cycle engines. In general, they have either greatly reduced octane requirements (PROCO) or no octane requirements at all (TCP), depending on the relative timing of fuel injection and of ignition. The multifuel capability of the TCP engine has been established,[45] and PROCO engines have been operated satisfactorily with blends of gasoline and alcohol or water, as well as with gasolines with modified volatility.[43] It appears that both the relative independence from octane requirements and the multifuel capability are enhanced by operating and design conditions which favor the emission of particulates and sacrifice of specific power. Therefore, some compromise may be necessary in order to minimize particulate emissions and to maintain power. However, even with such a compromise, it is clear that these engines have the capability of greatly broadening the range of acceptable fuel properties and therefore will allow the use of a larger portion of the petroleum crude or else will facilitate the use of alternate fuels. It is expected that unleaded fuels will still be required, since available emission data[42,43,45] indicate that oxidation catalysts will continue to be required.

5. Diesel Engines

The fuel requirements of diesel engines are well known. These engines do not have octane requirements, but rather require fuels with a high cetane number, a parameter related to the ease of igniting a fuel by compression heating. In general, the cetane number is complementary to the octane number: high octane fuels typically have low cetane numbers and vice versa. Historically, the relatively slow diesel engines used for stationary application have shown tolerance for a wide range of fuels, from coke oven gas to powdered coal, even though Rudolph Diesel's experiments with powdered coal were only partially successful. Efforts to use such extreme fuel choices are continuing,[44] even though liquid fuels are almost universally selected for automotive applications. Even in these applications, diesels tolerate a fair range of volatility.[46] However, compression ignition, as stated above, imposes a strong requirement for a high cetane number; this, in turn, can be a problem with some crudes.[47] These requirements are also an obstacle to the use of alcohols, which typically have very high octane numbers, but low cetane numbers; alcohols require the simultaneous use of a high cetane fuel, such as conventional diesel fuel, as a pilot to provide ignition.[48]

This author believes, however, that diesel engines will not become a major factor in passenger car propulsion. Although the diesel's disadvantages in terms of weight, cost, noise, startability, etc., may not be insurmountable, in this author's judgement the diesel will continue to suffer a significant emission penalty, particularly in terms of particulates. Even with the lowest particulate emission rates reported (about 0.1 g/km),[49,50] the increased atmospheric burden would be a cause of serious concern. It has been estimated that, if all the passenger cars in the Los Angeles basin were converted to diesels, the resulting increase in emissions would add about 30 mg/m^3 to the average atmospheric particulate loading in that area.[51] It is likely that EGR, which may be needed to reduce the emission of oxides of nitrogen from these engines, would aggravate particulate emissions.

6. Stirling and Gas Turbine Engines

A recent comprehensive study[52] identified the Stirling and gas turbine engines as the choices for the future. However, it seems unlikely that these engines will be a major factor in automotive propulsion before the 1990's.[42] The advantages identified for the Stirling and possibly for the gas turbine engine are that they combine the potential for very low emissions, without the need for exhaust aftertreatment, with excellent fuel economy.[53, 54] In the case of the gas turbine, the achievement of the latter requires high operating temperatures, such as can be achieved by the use of a ceramic turbine power patch.[53] A discussion of the remaining development required for these engines is beyond the scope of this paper. The main issues have been enumerated previously.[42]

Possibly a significant advantage of these engines is their adaptability to a wide variety of fuels, without any restrictions as to volatility, octane, or cetane. Nevertheless, some limits may be expected in terms of contaminants. Thus, fuel nitrogen would be expected to increase the emissions of oxides of nitrogen[55]; other emission effects of fuel subsitution have been reported[56]; and, in any case, it is expected that combustor changes will be required to achieve high efficiency and low emissions with different fuels. It is not known yet what fuel composition requirements, if any, will be needed to achieve low emissions.

7. Concluding Remarks

The traditional automotive engines—Otto and diesel—will continue to be the dominant engines in automotive transportation for a long time. Their main fuel requirements are related to ignition: octane number for the former, cetane for the latter. In addition, the Otto engine requires either gaseous fuels or high volatility liquid fuels, while the diesel is best suited for liquid fuels with intermediate volatility. Additional restrictions are imposed by the emission control requirements. Nevertheless, both engines can be adapted to a wide variety of fuels. The main limitation in the future is likely to result not so much from the characteristics of these engines as from the economics of production of alternative fuels and their compatibility with existing storage and distribution systems.

Direct injection, stratified charge engines have less stringent octane requirements (none at all in some cases) as well as greater flexibility in fuel volatility. Since it is likely that these engines will still require oxidation catalysts, they will still require unleaded fuels. However, they may be adapted even more easily than the conventional engines to the use of alternative fuels such as alcohols as well as to wider boiling-range fuel.

Stirling and gas turbine engines have neither octane nor cetane requirements, as well as no volatility restrictions. Fuel composition requirements needed to achieve low emissions are still to be determined. It is possible that these engines can be adapted to the use of synthetic fuels such as may be obtained for instance from coal or oil shale, with reduced refining requirements compared to conventional engines.

Probably for the better part of the next two decades the main automotive fuel will continue to have properties similar to gasoline, no matter what alternative source is used as a replacement for petroleum. Although engines could be adapted easily to the use of other fuels such as alcohols, the advantages of such substitutions have still to be established.

B.   Alternative Fuels for Buses, Trucks, and Off-Highway Equipment

Thomas C. Young

This paper is based on a study made by the Alternate Fuels Committee of the Engine Manufacturers Association (EMA). Members of the Association manufacture internal combustion engines, both gasoline and diesel, for nonautomobile uses. This study, limited to portable fuels suitable for use in these engines, was undertaken with the following objectives:

(a) To determine physical and other changes likely to occur in petroleum-based fuels and assess their impact on engine manufacturers.

(b) To identify alternate fuels most likely to supplement or replace petroleum fuels and assess their impact on engine manufacturers.

(c) To recommend research and development programs to better identify or reduce the impact of the above changes.

(d) To establish a position on current and alternate fuels so that EMA can make positive contributions to possible legislation or regulations.

(e) To assess the supply/demand situation.

To put the truck, bus, and construction vehicle sectors into perspective, it should be noted that they differ substantially from the automobile sector. First, these are commercial vehicles, purchased for the most part as investments to help provide profits. Second, while the total of new units numbers only between 350,000 and 500,000 per year (as opposed to many millions for automobiles), the cost per unit can go to $100,000. These two factors (function and cost) weigh heavily in any consideration of alternative fuels. It is obvious that fuel economy is and will continue to be of major concern in the purchase of these products, and is therefore of major concern to the engine manufacturer.

The impact of these vehicles on the total fuel situation is considerable (not to mention their impact on commerce generally: in 1975, trucks carried 20% of U.S. freight transportation, better than 535 billion ton-miles—and this could double by 1995). Diesel fuel, the primary fuel for these vehicles, accounted for 25.5% of the total distillate fuel used in 1975 (266 million barrels) and 4.5% of the total petroleum demand (Table 20).

The criteria used in judging the alternative fuels were as follows: (a) fuel availability, (b) changing fuel parameters, (c) technological state of the art, (d) costs—investment and operating, and (e) energy conversion efficiency.

Consideration of all major portable fuels and their sources included estimating each one's overall energy conversion efficiency across the

Table 20   1975 U.S. Distillate Fuel Oil Usage[a]

	Million Barrels/ Year	% of Total Petroleum	% of Total Distillate Fuel Oil
Total Petroleum Demand	5,957	100.0	...
Total Distillate Fuel Oil Demand	1,043	17.5	100.0
Distillate Fuel Oil Uses[b]			
Heating Oil	487	8.2	46.7
Diesel Fuel (On- and Off-Highway)	266	4.5	25.5
Diesel Fuel (Railroad)	93	1.5	8.9
Electric Utility	65	1.1	6.2
Industrial	64	1.1	6.2
Vessels (Bunkering)	26	0.4	2.5
Military	18	0.3	1.7
Oil Company	14	0.2	1.3
Miscellaneous	10	0.2	1.0

[a] Source: *API Basic Petroleum Data Book,* Petroleum Industry Statistics (taken from U.S. Bureau of Mines, Sales of Fuel Oil and Kerosene).
[b] Does not include distillate used for kerosene and jet fuel.

entire conversion process, from its raw state to its ultimate use in the vehicle. For example, shale oil was followed through the following conversion paths: above-ground retorting, in situ, and kerogen. Coal was examined as a solid fuel, liquefied, converted to methanol, converted to electricity, and converted to hydrogen (both from the liquid and hydride). Similarly, all major sources and processes were reviewed and analyzed. The changes in fuel parameters will be different in the short and long terms.

1. The Short Term

The short-term analyses are based on the use of petroleum as the fuel source. The production of diesel fuel and of gasoline are interrelated. The petroleum industry has traditionally provided gasoline to meet the octane requirements of spark-ignition engines through refining techniques and the use of such additivies as tetraethyl lead. The introduction of unleaded gasoline in 1973 brought along the additive methylcyclopentadienl manganese tricarbonyl (MMT) in 1975, to stretch out the supply of so-called clear octane gasoline. Both additives are now under fire from the EPA. Lead content of gasoline must be reduced from 1.8 grams per gallon to 0.5 by October 1979. (This ruling is being contested in court, but it is expected to stand.) And MMT shows signs of plugging catalytic converters, thereby increasing hydrocarbon emissions and lowering fuel economy. Because of the lead and MMT regulations, the petroleum industry will have to increase octane number through additional refining, which will be costly in both capital and energy. In the very short term, until about 1980, the anticipated effects of these rulings are either a reduction in gasoline availability by about 7%, or a lowering of fuel quality. If the quality were lowered, i.e., the octane number reduced, the thermal efficiency of spark-ignition engines would drop correspondingly. This in turn would increase the demand for gasoline, adding further pressure on the U.S. refining capacity. Beyond 1980, the refining industry would have to invest about $2 billion to upgrade quality.

What does that do to diesel fuel, at least for the short term? The more severe cracking to increase production of higher octane clear gasoline will reduce the cetane number of the diesel fuel by-product. With our increasing dependence on foreign crude oil with its higher aromatic content (which adversely affects diesel fuels), we can reasonably expect a progressive deterioration in cetane number. Indeed, we now feel that, as we get closer to 1990, we can expect ever-increasing proportions of minimum-specification U.S.-produced diesel fuel; i.e., down to the ASTM minimum of 40 cetane.

Table 21   Typical Diesel Fuel Oil Specifications:
U.S. Versus Europe (1977)

	U.S.-Produced	Non-U.S.
Gravity, API	30 to 36 (31)[a]	36 to 47 (38)
Cetane Number	39 to 60 (47)	55 to 65
Distillation Range, °F	350 to 360	350 to 700
Cloud Point, °F	− 16 to + 16 (5)	− 5 to 20
Pour Point, °F	− 40 to + 5 (− 10)	− 5 to 15

[a]( ) = typical value.

Table 21 shows the differences in diesel fuel specifications in Europe and the U.S., including typical values. The differences in cetane numbers and cloud points are primarily due to refining procedures. U.S. refineries crack every residual to increase gasoline yield. This practice is not as prevalent abroad. Traditionally, in winter No. 1 diesel fuel or kerosene has been added to No. 2 diesel fuel to achieve the necessary filterability. Now, with increasing importation of European No. 2 fuel with its consistently higher cloud points, plus a trend toward higher end-points for U.S.-produced fuels and the diminishing availability of No. 1 fuels, there is a greater likelihood of operational problems in winter.

2.   The Long Term

The long-term prospect is that alternatives will replace the diminishing petroleum supply and that coal and oil shale are the most likely candidates. They are in abundant supply and are capable of being converted to fuels very similar to those in use today. Fuels from shale oil can be produced with overall efficiencies of 50-60%; liquid hydrocarbons from coal, about 50%; and methanol from coal and water, near 40%. Costs will be higher due to technological and environmental constraints (Table 22), but much of the technology is known.

There exist little test data showing the impact of the alternative fuels on engines. We can foresee some difficulties, however, and these should be explored:

(a) Lowering the cetane will bring problems with cold starting, noise, smoke, and emissions.

(b) Heavier crudes may cause an increase in carbon deposits, smoke, and emissions.

(c) The emergence of the alcohols, with their lower bulk energy content and low lubricity (for fuel-lubricated injection pumps in diesels), could require the development of new fuel systems.

(d) Decreasing metallic additives in gasoline can affect engine

Table 22   Alternate Fuel Constraints

Fuels	Constraints
Oil from Shale	Extraction technology; water at site; environmental; cost.
Liquid Hydrocarbons from Coal	Conversion technology; energy efficiency; environmental.
Methanol Fuel from Coal	Energy efficiency; environmental.
Ethanol from Biomass	Energy efficiency; limited production rate.
Improved Solid Fuels	Technology of conversion and engine utilization.
Hydrogen	Distribution and storage.

durability; decreasing the octane number can affect thermal efficiency.

(e) Higher cloud points in diesel fuel can increase filter blockage.

The study's conclusions regarding alternative fuels for nonautomobile road vehicles may be summarized as follows:

(a) It will be necessary to push for development of alternative liquid fuels beyond 1990, and there will be a critical availability of fuels in the years 2000 to 2025.

(b) It is obvious that imports will remain a major source of energy supply for the United States.

(c) Companies involved with internal combustion engines should monitor changing fuel specifications as they impact upon engine performance and consider appropriate research and development to meet such changes. Suppliers may also have to adapt engines to specialized areas like Brazil, where the use of alcohol fuels may be of considerable merit on a broad basis.

(d) Alternative fuels based on coal or shale provide the most desirable fuels for the short term for portable internal combustion engine use.

(e) We need to develop a coherent U.S. national energy policy which would include the use of conservation of efforts; would free markets (i.e., remove price controls and subsidies) so that economics can allocate uses of conventional fuels; and would consider environmental tradeoffs and appropriate research and development.

(f) For the long term, we must continue to develop nuclear and solar energy sources.

During the course of this study there was a natural desire to attempt to define research gaps and establish such gaps in order of priority. However, the problems are so massive and the time frame so short that several research and development avenues should be pursued concurrently. Limiting research and development to one or two major avenues introduces too great a risk.

C. Outlook for Electric Road Vehicles

William Hamilton
General Research Corporation
Santa Barbara, California

1. Abstract

Electric vehicles in very large numbers could be recharged from coal or nuclear facilities now available or projected at U.S. utilities. Resultant savings of petroleum could be correspondingly large. With future batteries, the driving range of electric cars will become adequate for most automotive travel. Unless gasoline prices increase drastically in relation to oil prices or government intervenes directly, however, relatively few electric cars are likely to be sold in this century, because they generally will cost more and do less than conventional cars.

2. Introduction

Electric vehicles offer what may be the ultimate versatility for using alternative fuels. They can be operated with high efficiency from petroleum, natural gas, coal, hydropower, nuclear reactors, geothermal steam, solar cells, or any other source of energy to be used by electric utilities. Electric vehicles offer other advantages as well: they can be exceptionally quiet, reliable, long-lived, and maintenance-free; and they emit no air pollutants. These advantages together with their potential savings of petroleum have led government and industry around the world to undertake developments of improved electric vehicles and improved storage batteries for them. The aim of these programs is to minimize the disadvantages of electric cars which have long denied them a sizable share of the vehicular market: limited range and higher cost.

The effects of large-scale use of electric automobiles on energy, the environment, and the economy in the United States have been projected in a recently-completed study for the Department of Energy.[57] This study provides projections to be presented in this paper for availability of energy for recharging electric vehicles and the fuels from which it would be derived. Also discussed here are the probable applicability and sales of electric vehicles and the influential role likely to be played by government in their future.

3. Availability of Energy for Recharge

The U.S. utility industry is bigger and probably even more difficult to change than the U.S. automobile industry. Utility sales are approximately equal to those of motor vehicle manufacturers, but utility capital investment is over six times larger. Moreover, the utility industry may be the most regulated large industry in the United States.

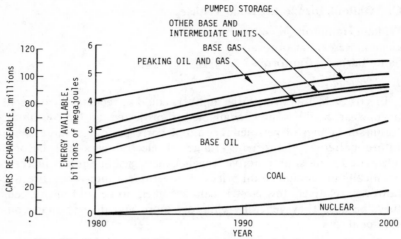

Fig. 23   Energy available on peak days for electric car recharge

Major changes in its facilities would therefore be most difficult to make.

No changes are needed, however, in order to accommodate recharging of enormous numbers of electric vehicles. Utility capacity is sufficient to meet the peak demand of each year; average demand is much less, and even on the day of peak demand, a utility's load is likely to drop to little over half the peak by the early hours of the next morning. In 1975, utilities actually generated about five times as much electricity as would have been required to electrify all automobiles in the United States.

The energy which could be generated in the United States to recharge electric cars on the peak-demand day of the year, when idle facilities are fewest, is shown in Figure 23. The energy available on most other days would be far greater. The projection of Figure 23 is based on detailed projections of supply and demand reported by electric utilities to the Federal Power Commission and the Edison Electric Institute. It was calculated by a large computer model which projected for each utility the capacity available by type of fuel, and its use to meet projected demand in each hour of a future year. Since the electric utility industry projects more rapid growth than that expected for travel by motor vehicles, the percentage of vehicles which could be electrified within available capacity increases in future years. Because increased capacity is almost entirely expected to be nuclear or coal, the relative importance of petroleum declines continuously in this century. Future electric cars will require about 1 MJ of recharge energy per kilometer of travel, or a little less than 50 MJ for the average day's driving. This leads to the numbers of cars which could be recharged shown in Figure 23.

To make effective use of energy available for recharging, selective load control would probably be necessary. This technological innovation, becoming common in Europe and already in use at a few U.S. utilities, allows utilities to exercise remote control over such interruptable loads as electric water heating and space heating. It could ensure that battery chargers did not increase peak loads unacceptably, and that charging was done at times when the most desirable fuels could be used for generating needed electricity.

Selective load control could also be used to reduce peak loads and planned expansion of utilities. In this event, of course, less generating capability would be available for recharging electric vehicles than is projected in Figure 23. There are many other uncertainties as well in this projection, notably whether demand growth will remain much lower than in years prior to the 1973-4 oil shortage, and whether nuclear plant construction will proceed at planned rates.

4. Use of Fuels and Energy by Electric Cars

Electric cars of the future should be much superior to those presently available, largely because of extensive governmental support for improving batteries (historically the weak link in electric vehicles) and vehicle designs. In a relatively few years, there is an excellent prospect that both improved lead-acid and nickel-zinc batteries will become available with the low initial costs and long service lifetimes required to make them practical. Whereas today's lead-acid batteries produce only 100 kilojoules of output per kilogram, the advanced lead-acid batteries should reach 170 kilojoules per kilogram, and the nickel-zinc batteries should reach 280 kilojoules per kilogram. By the 1990's, very advanced molten-salt batteries may achieve well over 400 kJ/kg.

The potential offered by the improved batteries is considerable: practical ranges in urban driving of 150 and 250 kilometers, respectively, with ranges on the freeway which are even greater. This is far in excess of the average daily auto use in the United States, which is about 50 kilometers. A future electric car with these capabilities, depending on the battery used, is compared with a conventional subcompact car in Figure 24. The portions of the body accommodating the passengers in the electric and internal-combustion engine (ICE) car are identical. The extra weight of the electric car, about 85%, results from the large battery required. This extra weight includes additional structure to support the battery, and additional propulsion to maintain performance adequate for safe entry into freeways. Because of its extra weight even without battery, the electric

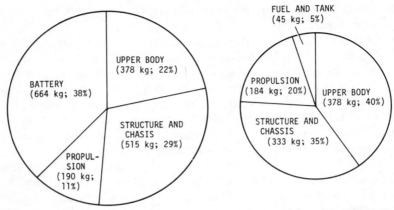

"IMPROVED" ELECTRIC SUBCOMPACT

- 1747 kg CURB WEIGHT
- 43.3 kW PEAK POWER AVAILABLE (23 W PER kg OF TEST WEIGHT)
- 250 km RANGE (NICKEL-ZINC BATTERY)
- 150 km RANGE (LEAD-ACID BATTERY)

"WEIGHT CONSCIOUS" ICE SUBCOMPACT

- 940 kg CURB WEIGHT
- 53 kW CONTINUOUS POWER AVAILABLE (49.3 W PER kg OF TEST WEIGHT)
- UNLIMITED RANGE (WITH REFUELING)

Fig. 24   A comparison of electric and conventional cars

car is unlikely to be cheaper than the conventional car. Moreover, energy use of automobiles is proportional, in the first approximation, to weight, so the electric car is at a substantial initial disadvantage in that its weight is relatively high.

Nevertheless, electric cars can be competitive or even superior in required fossil fuel per kilometer of travel. In Figure 25, projected fuel requirements of conventional and electric cars are shown under the assumption that all energy for operation is derived solely from petroleum or solely from coal. Past requirements are also shown, in terms of petroleum only, for average cars on the road in 1955-75. The future requirements are shown for the electric cars of Figure 24, for lighter electric cars with smaller batteries and ranges of 100 or 150 km, and for projected conventional cars fueled with gasoline obtained either from petroleum or coal. Production of synthetic gasoline from coal is relatively inefficient. Electric cars, however, use energy from either oil or coal with approximately equal efficiency. While electric cars are at best competitive with conventional cars if petroleum is the primary fuel, they are far superior if coal is to be the source of energy for transportation.

5.   Savings of Petroleum Due to Electric Cars

If electric cars replace conventional cars, petroleum used to make gasoline will decrease in proportion. Relatively little oil will be used to

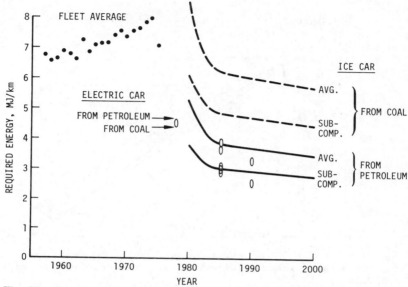

Fig. 25   Energy required from resources of coal or petroleum for urban driving (new cars)

generate recharge electricity, however, given the expanding capability of utilities to use coal and nuclear power. Projected use of petroleum in 2000 for conventional cars, for the electric cars which might replace them, and for all cars of both kinds is shown in Figure 26. A band of usage levels is shown for electric cars because oil use depends on where the electric cars are introduced. Use of oil for recharge would fall at the upper edge of the band if electric cars were distributed uniformly to utility customers in the United States. If they were first distributed in regions where utilities use the least petroleum for generating recharge energy, however, oil use would fall at the lower edge of the band. A band of uses is also shown for conventional cars; the lower edge shows usage if average cars are replaced by electric cars, while the upper edge shows usage if efficient subcompacts are first replaced. In 1990, almost 40% of U.S. automobiles could be replaced by electric cars with virtually no use of petroleum for recharge, if they were distributed in the most favorable areas. In 1980 and 2000, corresponding figures are 30% and 60%.

Regional differences in reliance on petroleum for recharging are suggested in Figure 27. This figure shows the mix of fuels which would be used in each of the nine regions of the National Electric Reliability Council if all cars were electrified. Reliance on petroleum would be least in central states with coal available, and greatest in Southern

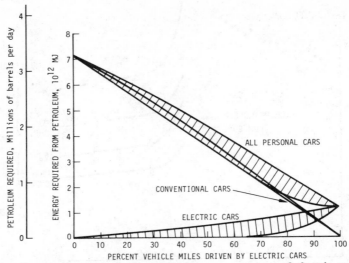

Fig. 26   Required petroleum for personal cars versus use of electric cars: year 2000

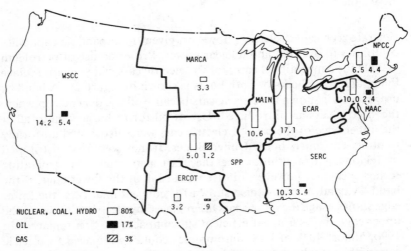

Fig. 27   Fuel mix by region for electrifying all cars in the year 2000

California and in New England. With 100% electrification, capacity would be insufficient on occasional days to recharge all cars fully. Even if travel were curtailed accordingly, only a few percent would be sacrificed. This would seldom be necessary, however, because most electric cars would have sufficient capability to accomplish three or four days of average driving even if they were not recharged at all.

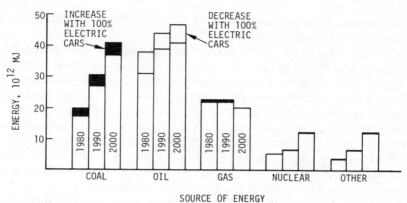

Fig. 28 **National annual use of energy with and without electric cars**

Projected effects of electric cars on energy use are shown in Figure 28. Since electric cars are no more efficient than conventional cars fueled from petroleum, what is saved in petroleum must be made up from other sources. Coal would be the primary source of this energy, with only a modest part supplied from nuclear plants. By themselves, it is clear that electric cars cannot solve the U.S. petroleum problem: today, automobiles consume only about a quarter of petroleum used in the United States, and even this figure is expected to decline as a result of improving automotive fuel economy.

6. Outlook for Electric Car

Given continued ready availability of gasoline and no intervention by the government, electric cars are unlikely to capture as much as 10% of the U.S. automobile market by the end of this century. The reason is simply that they will continue to offer less capability at higher cost than conventional cars, even if major improvements in batteries are achieved.

Both the higher cost and the lower capability are due to the battery itself, which adds weight and expense unnecessary in a conventional car and at the same time limits driving range. In purchase decisions made by motorists, higher cost and lower capability are likely to be far more important than independence of the gasoline pump, or the infinitesimal contributions to air quality and petroleum conservation which a single purchase of an electric car represents. For the user, the convenience and reliability of the electric car will probably be a more important advantage. Convenient home recharging will replace visits to service stations for fuel, and the inherent simplicity, longevity, and reliability of electric motors and controllers will eliminate the great majority of trips to the garage for maintenance and repair. These advantages, however, seem unlikely to offset the disadvantages of higher cost and lower capability.

Though electric cars seem unlikely to sell well, they will probably become capable of most of the driving U.S. motorists now require. Travel surveys indicate that urban secondary cars at multidriver households travel less than 75 kilometers on 95% of driving days. A car of this range can be built today; it could be applied as a secondary car with little sacrifice of travel. Urban cars at one-driver households ("only cars") are driven less than 150 kilometers on 95% of days, and urban primary cars at multidriver households are driven under 220 kilometers on 95% of days. Improved batteries should make these longer ranges achievable in mass-produced automobiles before 1990. With these increases, the applicability of electric cars would increase greatly, as shown in Figure 29. Separate curves are shown for cars at single-family housing units and at multifamily housing units. Overnight charging would be easier at single-family units; in multi-family units, providing individually metered, high-capacity electric outlets in lots and parking garages could be much more expensive and difficult to arrange. A band is shown for each type of housing unit corresponding to units which probably do not have off-street parking

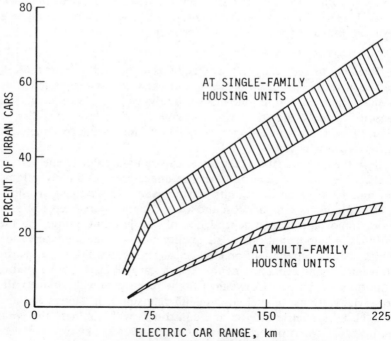

Fig. 29   Applicability of electric cars in urban areas

and therefore would have no place for recharging an electric car overnight. The data are poor, but a substantial minority of cars are apparently parked on the street overnight.

7.   *Outlook for Electric Trucks and Vans*

Electric vehicles excel in stop-start service, primarily because conventional vehicles perform poorly in it. Frequent periods of idling and frequent accelerations through the gears can drastically reduce both fuel economy and life of conventional vehicles. Electric vehicles, in contrast, show little or no impairment. Accordingly, they are likely to become economicaly competitive for such tasks as mail delivery, light urban goods delivery, meter reading, and the like long before they are competitive in passenger car service, where longer trips at higher speeds are relatively common. A major result of the U.S. government's demonstration of electric vehicles may be to popularize their use in these applications.

No detailed surveys have been made of the number of conventional vehicles in stop-start service which might cost-effectively be supplanted by electric vehicles. It does not, however, appear to be more than a percent or two of all U.S. road vehicles.

For trucks which are driven longer distances at higher speed, no batteries now in prospect offer sufficient capability for reasonable electrification. Most farm trucks, long-haul trucks, and many trucks in urban areas are thus beyond the apparent reach of electrification.

Beyond these commercial trucks there remain a large number of trucks and vans which are personal vehicles. Neither the precise number nor the usage of these vehicles is well documented. In the past 20 years, truck and van sales have increased from 12% to 24% of all motor vehicle sales in the U.S. with a total of over three million vehicles in 1976. Almost 75% of all trucks are pickups and vans, and a great many are used primarily as personal vehicles. For urban personal passenger transportation, electrification would be reasonable with the improved batteries hoped for by 1990. It is probable, however, that these trucks and vans are also used at least occasionally for long recreational trips, which may include the carrying of recreational equipment such as truck campers or the towing of trailers. Battery-electric drive with sufficient capability for these trips seems unlikely in this century.

Sales of four-wheel-drive trucks and motor homes have been rising rapidly in recent years. In 1976, over a quarter of a million motor homes were sold, 25 times the number sold in 1966. Electrification of such vehicles would require batteries far beyond anything now considered likely. In an era of scarce fuels, however, they may once again become as infrequent as they were a decade or two ago.

8. *Critical Factors for the Future*

The relative costs of gasoline and electricity express the basic free-market incentive for motorists to purchase electric rather than conventional vehicles. At present, gasoline prices in the U.S. are as low as they were 25 years ago, in constant dollars, and are only half or a third of prices in most other industrialized nations. Increases to $1-$2 per gallon, depending on battery technology, would make shorter-range electric cars cost-competitive with ICE cars. The disadvantage of limited range would remain, but further increases in the price of gasoline relative to that of electricity could induce many motorists to make the switch.

Whether this will happen depends largely on the U.S. government, which now actively holds down the price of gasoline through direct regulation. Indeed, the entire future of electric cars may rest with government action. The situation is analogous to that of automotive air pollution and fuel economy. Individual motorists did not generally choose low-pollution or fuel-efficient cars in the free market. Collectively, however, through government regulation the people of the United States have required the use of both clean and efficient cars in order to obtain the collective benefits they bring. Use of electric cars might thus be similarly fostered, for the same reasons. The extra cost and loss of mobility would be very much greater, however, than in the case of low-pollution and fuel-efficient cars. Whether it is justified will be debated with increasing frequency if U.S. dependence on foreign crude oil continues to increase or its balance-of-payments deficit remains very high.

D. Commentary by Symposium Participants

(a) In a fleet test run by SOHIO, cars were run on alcohol blends with no special adjustment. The only reported difficulty in drivability was due to a tendency toward vapor lock. The car would have difficulty in restarting (but not initial starting) as readily as it should. Other problems, with "reasonable" amounts of alcohol, are not significant enough to bother the average driver. Those reasonable amounts vary, depending on the alcohol. At 20% methanol there would be problems. At 20% ethanol there don't seem to be any. This is borne out by the day-to-day experience in Brazil, where automobile fuel varies from straight gasoline as to much as 20% ethanol. To be sure, the cars are carbureted rich in order to minimize the effect of the leaning out of the fuel when the ethanol is added. While the addition of up to 20% ethanol to gasoline would be acceptable to most people in terms of drivability, it would have an effect on emissions that might not be acceptable to the EPA. Burning the mixture in a richly carbureted engine (as in the older cars) would result in a lessening of CO emissions, but an increase of NO_x; in an engine carbureted lean (as in recent engines) there would be a decrease in NO_x but an increase in the hydrocarbons. If you use a three-way catalyst system, where the proper stoichiometry is critical—usually just a very little bit richer —you would have to have automatic compensation with a closed-loop air-fuel ratio control system in order to keep the emissions from getting even worse. If the country wanted to go with a gasoline-ethanol blend at minimal expense in drivability, a small change in the carburetor to make the engine run richer would be necessary. This would have to be preceded by a change in the law. At present such a carburetor change would be illegal.

(b) *Question:* If one were to take crude shale oil, split it at 50% and get it down to a boiling range of 750°F, and have 2% nitrogen, could that fuel be used directly in a DISC engine, that is, without hydrocracking or hydrotreating?

Answer: Both yes, with the Texaco TCP, because it has no octane requirements; and no, with the Ford PROCO, because it does. The real difficulty here would be the NO_x emissions. When it is fuel nitrogen that is contributing to the NO_x emissions, then the low-temperature combustion in the DISC engines actually makes things worse. To avoid this problem, one would have to remove the nitrogen from the fuel, a very expensive step. A catalytic muffler would not do it, since it does not remove oxides of nitrogen under highly oxidizing conditions—and DISC engines perform most economically when there is an excess of air.

(c) *Question:* Don't the particulate regulations in effect restrict how high an end-point one can run on a DISC-engine fuel? That is, when using a DISC engine, the higher the end-point, the greater the particulate generation. And yet if one were to lower the end-point, then the cetane would suffer.

Answer: In the PROCO engine, the interval between injection and ignition—which causes the disadvantage of the octane requirement—is an advantage here, in that it reduces the tendency for particulates to form, even with a higher end-point.

(d) *Question:* Can you quantify the environmental benefits of electric cars?

Answer: In the mid-60's, four bills were introduced in Congress to promote electric cars specifically for benefits in air quality. In the intervening years, emissions from conventional cars have been cleaned up to an extent more far-reaching than that of any other single source of air pollution, and this cleaning-up will continue into the mid-80's. So the projected environmental benefits in changing to electric do not look as dramatic now as they did in 1965. But they are still real and should not be minimized. Automobiles at present account for between

10% and 50% of emission pollutants in urban air. (The percentage depends on the pollutant.) Those percentages would be the very most that could be removed if the switch were made. However, these would be offset by the emissions from whatever fuel is burned in the same air basin in order to recharge the cars. Overall, electric over conventional cars would reduce urban air pollution substantially—anywhere from 10% to 30%. There are other environmental benefits. Electric cars are quiet. For example, very little can be done to combat the acceleration noise of conventional cars. That is almost no problem with the electrics. Tire noise in cruise, however, would be about the same for both. Electric cars do away with the need for disposing of crankcase oil that so often is drained and dumped in sewers and storm-drains by do-it-yourselfers. At present, however, in the consideration of electric cars, these advantages seem to be of lesser priority than that of the saving of petroleum.

(e) *Question:* What are the expectations from the current federal electric-car demonstration?

Answer: The DOE program's aim is to develop and demonstrate the technology, which in turn will put thousands of electric vehicles on the road by 1985. If we assume that the technology will be successful—and we are assuming that—then there is a good possibility that the program will result in as many as 5% electric vehicles on the road by the 1990's. While this would not save a huge amount of petroleum—given the size of the total automobile market—it would still mean the mass production of hundreds of thousands of units per year, and billions of dollars in sales. This would in effect be a new "industry," and a tremendous accomplishment comparable in its effectiveness as a government program only with the nuclear power program, whose budget was orders of magnitude larger.

(f) The Brazil experience, according to Volkswagen, has been that cars of all manufacturers are running fairly well on a 20% ethanol/gasoline mixture—that is, without any major complaints. The leaning out of the mixture due to the alcohol brings the CO emissions down pretty well. Work is also being done to run cars on a 95% ethanol/5% water mixture, and about 500 research cars in Sao Paulo are already doing so. With some backing from the Brazilian government, work could go ahead to produce a few cars specifically to run on this "pure" ethanol. Major changes in the engines would be needed: for example, more heat to the manifold, better mixture distribution, different spark plugs and spark timing, and a higher compression ratio. At cold temperatures (it gets down to about 5°C in Brazil), a device in the car injects a small amount of gasoline to get the car started. Experiments indicate that an additive to the idling mixture will eliminate the need for this device.

(g) *Question:* Are there possible incompatibilities between existing engine lubricants and engines running on the ethanol blend?

Answer: While such incompatibilities cannot be ruled out, so far there has been no evidence of them.

(h) *Question:* Did any materials at all in the VW fuel systems have to be changed in order for the cars to operate satisfactorily with the blend?

Answer: So far, none.

(i) *Question:* Can we learn from Brazilian manufacturers how to overcome the problem of water in the system when we go to an ethanol-gasoline blend so that there would be no drivability problem?

Answer: The problem of water in the blend is one for the petroleum companies to solve, not for the manufacturers.

(j) In a recent emissions test, a Brazilian car was run on straight ethanol. The main problem was excessive emission of aldehydes. There were no unique problems with the regulated emissions, and if one wanted to use the fuel in such a manner as to meet current regulations, one could. But then one would not be taking advantage of ethanol's potential for greater fuel economy. And if one were to go after that, one would go with a lower exhaust temperature, which would present new difficulties with hydrocarbon emission. This would call for some additional ingenuity in the catalyst system.

(k) *Question:* Have any real tests been run comparing a hydride-storage hydrogen vehicle with an electric vehicle for limited-performance use?

Answer: As far as anyone present knows, only paper studies have been done. They indicate that, of the weight penalty each of them suffers in comparison to gasoline, the penalty is perhaps slightly less with the hydrogen vehicle. The real problem between them is not the vehicles themselves, but the distribution systems. The electrical distribution system, costly though it may be, is already in place. A hydrogen distribution system is not. If it were, the hydride car would appear to have some advantages over the electric. One study, which analyzed the two genres of vehicle across the system—i.e., a comparison that included costs and efficiencies of preparing and distributing the energy—determined that the total efficiency was 25% better for the hydrogen car than it was for the electric.

(l) Studies of the impact of converting to electric vehicles were limited to the use of the electric power system as it is planned by the electric utilities. Since there are no present plans to install breeder reactors and fusion reactors, the quantitative analyst could not consider them. Similarly, solar electric power is coming along, and one might speculate as to how it can be a source of energy for electric vehicles. But again, utilities have no plans in place for this; it too was not considered.

(m) Some long-range predictions regarding the improvements and efficiency that we might expect in the automobile engine are of interest. These, of course, are guesses. There is room, in the conventional engine, for about 10% improvement in efficiency. First there are changes that can be made in the combustion-chamber design by using a somewhat more complicated arrangement of valves. It would cost more to manufacture, but the efficiency would improve by several percent. Second, a significant jump in efficiency could be made by making the engine smaller. (One way to do this is simply to sacrifice performance.) Third, if we can learn to control the production of NO_x in the engine well enough to avoid going to a three-way catalyst, we would avoid the loss of efficiency that the catalyst brings along. As for other automobile engines, the PROCO, primarily because of its lean operation, holds a potential 25% improvement in fuel economy over the conventional engine. The Stirling, which began by showing a 30% improvement over conventional engines, has not held on to that figure because of efficiency gains in conventional engines. The Stirling would need additional improvement to get back to 30%. The gas-turbine engine is an open question. If we can learn to operate it at the very high temperatures it needs for maximum efficiency, it would achieve a 30% gain in fuel economy over the conventional engine. Of course in certain cases these figures will vary according to the fuel that is used. Gasoline from shale, unless it is treated, is relatively low-octane. Its effect on the efficiency of conventional engines would be negative. The alcohols, on the other hand, can be used more efficiently, and hydrogen still more. Their effects would be positive. In the DISC engines, these factors would be much less significant. In the Stirling and gas-turbine engines, as long as there is no excessive flame luminosity—which could be a problem with a coal liquefaction fuel, and which can burn the combustors—changes in fuels would have no effect at all on efficiency.

(n) *Question:* Would going way up in octane number and compression ratio be a way of getting greater efficiency from the conventional engine?

Answer: When the compression ratio gets beyond about 11 to 1, friction losses begin to cancel the gains in efficiency.

(o) *Question:* In going to higher octanes, the polynuclear aromatics tend to increase. Can this be avoided with catalysts?

Answer: Yes. Catalysts are extremely effective in eliminating all the polynuclear aromatics (which are not of great concern), except for those with lower molecular weights. This discussion of higher octanes, incidentally, should not be interpreted as advocating an octane race. With present fuels and present refining methods, pushing the octane higher would have a negative effect on energy availability; the increase in refining energy consumption would more than offset any increase in engine efficiency. However, with the alternative fuels coming in, there are possibilities for higher octanes.

(p) There is some concern that the baseline against which fuel economy is measured keeps moving. For example, measurements made by running an engine on a dynamometer may be *called* miles per gallon, but they are not. For that you must run a real test on a real vehicle on a real road. Likewise, alternative engines and/or fuels will require differences in automobile design, such as volume or weight changes, and since these will affect fuel economy, they should be programmed in. Likewise, since emission standards will assume increasing importance, and will therefore increasingly affect fuel economy, these two variables should be tracked simultaneously. Finally, since alternative fuels are not expected to be primary until the late 80's and early 90's, we have to be thinking *now* about what the conventional cars are going to be then. In short, the baseline problem is a very difficult one.

(q) The German government is in the process of deciding on an R&D program for alternate automotive fuel systems; the program, as it now stands, will run from 1979 through 1982. It is expected to be funded with up to 160 million DM (about $70 million). The program will be divided into four areas: (a) electric cars, (b) alcohol-fuel engines, including trucks and buses, (c) hydrogen-fueled engines, and (d) hybrid engines. All German manufacturers and refineries will participate.

VI.
NONROAD VEHICLE UTILIZATION

A. Outlook for Performance of Alternate Fuels in Marine Vessels

E.G. Barry
Mobil Research and Development Corporation
Paulsboro, New Jersey

1. Abstract

This paper presents an overview of the U.S. marine fuel bunkering market and its relation to steam and diesel propulsion equipment. Principal focus is on residual fuel, covering present and projected future variability and performance requirements for both diesel and steam applications. Key fuel property requirements are highlighted, especially as they affect alternate fuel composition. Although the outlook on a technical basis could be considered favorable for liquid type alternate fuels, the ultimate acceptance will also be based upon their economic development together with the cost of equipment modification and maintenance.

2. Marine Fuel Market

Currently, two types of fuels are used in vessel bunkering: distillate fuels marketed as Marine Gas Oil or Marine Diesel Fuel and residual fuel blends marketed as Marine or Light Marine Fuel Oil. Specific grade use will be discussed in subsequent sections. To provide perspective regarding the consumption of these fuels, current federal[58] and industry[59] statistics indicate that U.S. distillate and residual bunker fuels account for 0.4 and 1.8%, respectively, of the total annual petroleum energy consumed. These data, compared to other petroleum fuels, are on Table 23. Further, these data indicate that 1.5% of the demand for all distillate fuel and 8.9% of the demand for all residual fuel is for vessel bunkers. These values have been gradually increasing over the past years. The data on Figure 30 indicate this growth trend up to the 1975 demand for distillate bunkers of 26,138 MB/yr and for residual fuel bunkers of 96,673 MB/yr. Thus, the U.S. marine bunker market represents a small but important sector of petroleum energy consumers.

3. Marine Fuels

Today, the bunkering terminals offer the marine trade a wide variety of fuels. These range from low-viscosity light-colored Marine

Table 23   U.S. Petroleum Transportation Energy Consumption[58,59]

Fuel	% Based on Total Energy Consumed	% Based on Total[a] Fuel Type Consumed
LPG	0.4	35.8
Gasoline	41.9	100
Jet Fuel	4.6	100
Distillate	6.8	32.5
(Vessel Bunkering)	(0.4)	(1.9)
Residual	2.3	11.3
(Vessel Bunkering)	(1.8)	(8.9)
Total (for Transportation)	56	...

[a] Transportation and nontransportation.

Gas Oil to high-viscosity dark-colored Marine Fuel Oil blended to the customer's specific request. Most of the distillate-type bunkers (i.e., Marine Gas Oil and Marine Diesel) are used in high-speed and locomotive-type diesel equipment. Since these are covered in other papers, this discussion will deal with the residual-fuel-type bunkers and their use in diesel and steam equipment.

In the past, the Light Marine Fuel Oil grade nomenclature system was based on the Redwood No. 1 viscosity. The numerical grade identification was the actual Redwood viscosity or the time in seconds for a specified volume of fuel to pass through an orifice of fixed diameter. The Redwood No. 1 viscosity is now considered obsolete and no longer an Institute of Petroleum method of test. Thus, recognizing the desirability of changing to a more technically oriented viscosity measuring system and at a more significant temperature, the marine fuel industry has recently changed from Redwood No. 1 viscosity at 100°F to kinematic viscosity in centistokes at 50°C.

This new grade system, which will now be universally adopted by the international marine bunkering market, and several of the equivalent grades based on the old system are shown in Table 24.

The selection of the fuel viscosity grade to be bunkered rests with the customer, and he is naturally interested in obtaining optimum vessel performance at the lowest possible energy cost. But, as the vessel obtains fuel at various terminals, the fuel properties will change depending upon the type of crude oil being refined, or the fuel being imported into that area. Table 25 contains examples of LMFO 180 fuels from the Middle East, North Africa, and Venezuela. Although there are many other crudes, these bracket the range of characteristics very well. Also, these are typical of the large production volume crudes which affect the properties of marine fuel in broad geographic

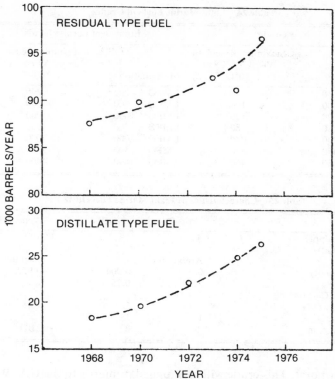

Fig. 30   U.S. vessel bunker fuel consumption trend

areas. The characteristics of the Middle East fuels are acceptable, although the sulfur content tends to be high. The Venezuelan fuel is marketed in the Caribbean and U.S. East Coast. It is high in vanadium and has a medium sulfur level. It is expected that these two basic fuels will continue to be typical of marine fuels. However, fuels may also be available from North and West African sources such as from Libyan and Nigerian crude oil. The typical North African fuel has a high pour point, making it difficult to handle in marine use. If it is to be applied as bunker fuel, it must be blended with other fuels to adjust the pour point to a manageable level.

In the future, both the North Sea and the North Slope of Alaska will supply substantial amounts of crude oil. Table 26 shows typical marine fuels from these sources. The North Sea based fuel is much like the North African: low in sulfur and vanadium. Thus, it will be a preferred fuel for inland use and very little of it is expected to appear in marine fuels. The Alaskan crude produces a more typical heavy fuel. It has a moderate sulfur content and moderate vanadium content

Table 24   Light Marine Fuel Oil Grades

New Grade		Equivalent Former Grade	
Designation	Maximum Kinematic Viscosity, CS at 50°C	Designation	Maximum Redwood No. 1 Viscosity at 100°F
LMFO 30	30	LMFO 200	200
LMFO 100	100	LMFO 800	800
LMFO 120	120	LMFO 1000	1000
LMFO 180	180	LMFO 1500	1500
LMFO 240	240	LMFO 2000	2000
LMFO 380	380	LMFO 3500	3500
LMFO 460	460	LMFO 4400	4400

Table 25   Characteristics of Light Marine Fuel Oil 180

	Mid-East	North Africa	Venezuela
% World Supply (Estimated)	44	18	10
Crude	Arab Light	Amal	Lagunillas
Specific Gravity	0.952	0.904	0.952
Sulfur, wt.%	3.1	0.25	2.1
Conradson Carbon Residue, wt. %	7.9	5.7	10.5
Pour Point, °C	13	38	− 18
Vanadium, ppm	26	2	210

and pour point. This crude will be supplied primarily to the U.S. West Coast refineries and it is expected that some will be used in marine fuel oil.

In addition to crude source, the refining techniques will also affect marine fuel properties. As the yield of gasoline and distillate fuel from crude is increased, more and more use will be made of catalytic cracking and in some cases thermal cracking or visbreaking.The effect of these changes is illustrated on Table 27. As cracked stocks are used, the specific gravity of the fuel increases, there will be little change in sulfur content, and a significant lowering of the pour point and an increase in the vanadium content will occur.

Currently, the fuels described in Tables 25-27 are marketed commercially and, provided they meet several performance properties discussed below, have been entirely satisfactory in both diesel and steam powered vessels.

4. Marine Propulsion Equipment Performance

The following discussion will review specific fuel performance requirements for diesel and steam plants. Available data on alternate fuels will be evaluated to determine their suitability in meeting these requirements.

Table 26   Marine Fuel Characteristics—New Sources

	North Sea	Alaska
Specific Gravity	0.916	0.963
Sulfur, wt. %	0.5	1.5
Conradson Carbon Residue, wt. %	3	8
Pour Point, °C	35	26
Vanadium, ppm	5	40

Table 27   Marine Fuel Characteristics—Refining Trends (Mid-East Crude)

	Conventionally Refined	Thermal Cracking Included	Thermal and Catalytic Cracking
Specific Gravity	0.952	0.985	0.995
Sulfur, wt. %	3.1	3.4	3.1
Pour Point, °C	13	−7	−4
Vanadium, ppm	26	50	60

Diesel Power. At the turn of the century, it became apparent that if the diesel engine were to become a commercial success a specific petroleum fraction, which ultimately became known as diesel fuel, would have to be used. The first such fuel was an all-distillate-type product. This became increasingly more expensive as the demand for gasoline increased and competed for the same portion of the crude oil barrel. Ideally the satisfactory use of a fuel similar in quality to steamship fuel would insure the continued growth of marine diesel power. This was accomplished through the early work of Lamb,[60] demonstrating that steamship-type fuel could be used in large-bore slow-speed marine engines provided the fuel was double centrifuged, first through a purifier centrifuge to remove water and gross contaminants and then through a clarifier-type centrifuge to remove more of the ash-forming material from the fuel. In addition, the viscosity of the fuel had to be adjusted by preheating to a value required for optimum drop formation; this normally ranges between 10 and 25 centistokes, depending upon the injector design. Successful application of centrifuging and viscosity control requirements has been further demonstrated for both slow[61,62] and medium speed engines[63-65] and has contributed to the fact that about 65% of the world's ocean-going tonnage currently is propelled by diesel engines.[66]

From this discussion, two performance requirements become apparent: viscosity control and centrifuging. Regarding viscosity, in order to achieve uniform injection conditions it is necessary to preheat

Table 28   Heavy Fuel Compatibility

Factors Causing Incompatibility
(a)   High molecular weight hydrocarbons in the heavy fuel component blended with...
(b)   distillate components which have low solubility for heavy asphaltic-type hydrocarbons can result in...
(c)   undissolved hydrocarbons separating from the mixture as incompatible sediment.

Table 29   Heavy Fuel Compatibility Test

Centrifuge—capable of 700 RCF[a] at the tips of the tubes
Centrifuge Tubes—100 ml cone-shaped type
Test Temperature—66°C (150°F)
Centrifuge Duration —3 hours
Sample Size—100 ml

[a] Relative centrifugal force.

the fuel oil. Thus, it can be concluded that as long as sufficient preheating facilities are available, the nominal viscosity of the fuel should have little effect. Therefore, fuels with viscosities higher than those grades listed on Table 24 could be used. This has been confirmed by the work of Grove[62] in which he concludes that, "In general, it has been experienced that the nominal viscosity of the fuel oil itself has very little effect on the lifetime of the engine components."

The need for centrifuging places requirements on the fuel density and incompatibility. To efficiently separate fuel from water contamination, a maximum fuel density is required to insure a minimum difference in density between the fuel and water. Work by Alfa-Laval[67] indicates that this results in a maximum fuel density of 0.984 at 15°C to insure good separation in their equipment.

The term incompatibility is used to describe those Light Fuel Oils wherein a portion of the high molecular weight hydrocarbons are not completely dissolved or suspended by the distillate. The factors causing incompatibility are shown on Table 28. Thus, when Light Fuel Oils are sold for diesel engines, precautions must be taken to insure that a compatible fuel will be supplied. To aid in this, the author's company has developed a centrifuge test which is described on Table 29. Our experience, reviewed below, indicates that this test can be useful in predicting fuel handling and combustion.

Table 30   Heavy Fuel Compatibility

Laboratory Engine Test Results
(a) Compatible Fuel ($<0.3\%$ Incompatible Sediment)—300 hours, no fuel-related problems.
(b) Incompatible Fuel (3-4% Incompatible Sediment)—after 200 hours, deterioration in fuel injector performance and decrease in combustion pressure, together with excessive deposits in the centrifuge.

In order to determine the effect of incompatible fuel on engine operation and establish a limiting specification value, a series of engine tests was made at Mobil's Paulsboro Research Laboratory and a survey of commercial fuels conducted.

In the laboratory study, a two-cylinder type "L" Bolnes marine diesel engine was run on a Light Marine Fuel Oil 150, containing up to 4% incompatible sediment. The run was scheduled for 300 hours or until fuel-related engine operating problems were encountered. Injector performance was monitored daily by means of the Mobil diesel diagnostic equipment. Combustion pressure patterns were obtained using a Kistler pickup, both initially and when operating problems were experienced or indicated by other instrumentation. Both the injector and combustion pressure patterns were obtained with the engine operating at full load.

Test results shown on Table 30 indicate that, when operating on this fuel, injector performance began to deteriorate after 200 hours, and there was a decrease in combustion pressure. In addition to encountering the injector plugging, the centrifuge was overloaded with sludge. In fact, the engine had to be shut down seven times during the course of the 300-hour run in order to clean out the centrifuge. In previous tests with Light Marine Fuel Oil 150 containing less than 0.3% incompatible sediment, no fuel-related problems were encountered.

In commercial operation, similar relationships were found between vessel performance and centrifuge sediment level as shown on Table 31. When the fuel used had a "low" sediment value (0.15%), no problems were encountered. A subsequent survey helped to define sediment limits, and it appears that, if the sediment value is maintained at 0.3% or lower, field problems can generally be avoided. Thus, this limit is currently recommended as a performance requirement for Light Marine Fuel Oils.

In addition to incompatibility sediment, there are other undesirable materials in Light Marine Fuel Oil which must be removed. Although

**Table 31 Fuel Performance Comparison
(Same Equipment—Different Bunkerings)**

Ship	Fuel Sample Incompatibility Sediment	Performance
A	0.15% vs 1.07%	No problem Centrifuge overloading
B	0.10% vs 0.50% vs 1.5%	No problem Excessive sediment Excessive sediment
C	0.15% vs 1.6%	No problem Purifier overloading, loss of power

centrifuging is the most prevalent cleaning technique, both filtration and homogenizers have also been used. Our work and that of others [68-70] has been aimed at evaluating these fuel cleaning techniques and what would account for the reported problems.

Various type filters are effective in removing large particles as shown on Table 32. But their effectiveness compared to the centrifuge is somewhat poorer in removing smaller particles ($< 5 \mu$m). If particles of this size contaminate heavy fuel, they can be abrasive and if not removed from the fuel could cause significant increases in cylinder liner and piston ring wear rate. Homogenizers, on the other hand, do not remove any of these abrasive particles and thus increased wear rates do result. Thus, properly maintained and operated centrifuges appear to offer the most effective fuel cleaning technique for marine engines.

In addition to unwanted and accidental pollutants, some investigators are suggesting that the various hydrocarbon types that comprise the bulk of the fuel may affect its performance. These are several of the reported problems: (a) high wear, (b) scuffing, (c) abnormally high combustion chamber temperatures, and (d) high radiation heat flux.

Wear rates, in particular, have become especially important in view of the additional requirements that the cylinder overhaul intervals be extended to beyond two years to reduce ship operating costs. This new requirement has even led to the development of techniques for real time wear measurements during the operation of laboratory [71] and commercial [72] diesel engines.

Table 32   Filtration System Effectiveness

	Fuel Before Treatment		Surface- Type Filter	Depth- Type Filter	Centrifuge
Asphalts, %	7.6		7.4	7.5	7.6
Oxidized Ash, %	0.036 (Fe, Na, Va)		0.06	0.05	0.04
Millipore Membrane, %					
>8	Av.	0.05	0.029	0.007	0.01
3	Av.	0.06	0.05	0.034	0.021
1.2	Av.	0.065	0.062	0.04	0.021
0.3	0.26	0.1	0.2	0.16	0.08
Total	0.435	0.275	0.296	0.241	0.132

As a result of this concern for wear as related to fuel properties, we are currently investigating the basic combustion properties of the various types of marine fuels shown on Tables 25-27. A turbocharged trunk-type laboratory Bolnes engine, instrumented with an engine combustion analyzer, is being used to measure (a) combustion delay, (b) rate of rise of combustion pressure, (c) duration of burning, (d) rate of heat release, and (e) other key combustion properties.

Preliminary data for a Venezuelan LMFO 150 are on Figure 31. The upper data plot is the fuel injection pressure at various engine crank angle degrees. From this, the actual point of fuel nozzle opening and the initiation as well as the duration of fuel injection are obtained. In the second data plot, measured combustion chamber pressure is recorded, again at various crank angle degrees. From this, the location of fuel ignition, the maximum or peak cylinder pressure, its location, and the rate of pressure rise can be obtained. The delay time between the initiation of fuel injection (upper plot) and combustion (lower plot) is one key measurement of the combustion quality of the fuel.

The bottom two plots contain both the rate of heat release (in kilocalories/second) from the burning fuel inside the combustion chamber as well as the total amount of heat released at various crank angle degrees. These data are obtained through an analog-type computer that measures various engine operating parameters, performs the proper thermodynamics, and provides an instantaneous data output.

In the rate of heat release plot, the time of ignition is more clearly defined (than in the pressure versus timing plot) as well as a measure of the rapidity of fuel burning. Note that as the exhaust valves open and hot gases leave the cylinder, there is a flow of heat energy out from the combustion chamber. The bottom plot of cumulative heat

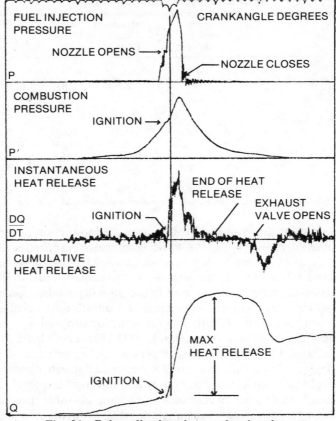

Fig. 31   Bolnes diesel engine combustion data

release indicates the total amount of heat available for work at specific crank angles. Some will pass through the cylinder walls to the coolant, part will be converted by expansion to useful mechanical work in turning the engine, and part will exit as hot exhaust gases, to be partially recovered as power for the turbocharger. From this latter plot, combustion initiation is evident as well as the smoothness of burning. Also, the time and completeness of burning prior to port opening can be measured.

Table 33 contains typical data from this test, including ignition delay and the location and magnitude of the peak combustion pressure. In addition, preliminary data from our heat release computer include the location of the maximum rate of heat release in the combustion cycle, the duration of fuel burning, and the amount of heat energy released as the piston reaches the top of its stroke or TDC.

Table 33   Bolnes Engine Combustion Studies—LMFO 150

	Typical Results
Basic Engine Timing	
Fuel Injection	14° BTC
Nozzle Opening	9.5° BTC
Ignition Delay	4° CA
Peak Pressure	875 psi
Location of Peak Pressure	9° ATC
Heat Energy Release	
Maximum Rate of Heat Release at...	8° ATC
Duration of Burning, °CA	72
% Heat Released at TDC	7

Obviously, all of these factors will relate to the ultimate efficiency of the engine. Thus with these devices, some insight into the key fuel variables that may affect engine combustion efficiency will be obtained.

Preliminary data comparing conventionally refined fuels (e.g., Table 25) and those containing thermally or catalytically cracked stocks (e.g., Table 27) indicate little difference in basic combustion parameters and engine efficiency.

Alternate Fuel/Diesel Power. From this discussion it can be seen that, as far as liquid-type alternate fuels are considered, there are definite density and incompatibility requirements for marine diesel applications to permit centrifuging and minimize wear. There are only minimal viscosity requirements although high pour points, above 35-40°C, could pose a handling problem without special heating facilities. Although short-term combustion studies would question the necessity for a minimum cetane number requirement, a long-term test would be needed to determine the fuel effects on economy, deposits, and wear.

Coal and coal/oil slurries could also be considered. However, prior to engine studies, an equipment design program is necessary to develop efficient techniques to handle and inject the fuel. In this regard, a patent has recently been issued to Sulzer Brothers Ltd. for a "Method and Apparatus for Operating an Internal Combustion Engine with Solid Fuel."[73] Another problem with coal will be wear. Many coals contain 7-14% ash, half of which is usually silica and 20-30% alumina, both highly abrasive minerals. Thus, while there may be sufficient time in a slow-speed marine diesel engine to effectively and efficiently burn the coal, the wear problem may preclude its use.

Steam Power. In a study entitled "All About Boilers,"[74] it is pointed out that "The thermal efficiency of the diesel engine makes this prime mover a strong contender. Diesels use less fuel per brake horsepower produced than any other kind of engine and they can burn marine heavy oil as well. But steam boilers should receive the attention. With the increasing demand for middle and heavy distillate it is only the steam boiler that can burn viscous and dirty Bunker C fuel oil, petroleum coke or coal." Ironically, the technical development that significantly contributed to the present-day acceptance of steam is judged by many manufacturers[75] to be the introduction of furnace waterwalls in place of firebrick. This reduction in maintenance somewhat offsets the higher operating fuel cost due to the lower efficiency. Thus, combining the ability to burn low-quality fuel[76] with reduced maintenance and more efficient modern marine systems (i.e., Very Advanced Propulsion turbine plant[77]) may indicate a future increase in interest in steam propulsion.

Even though the boiler can handle "low"-quality fuel the atomizing viscosity is still one of the most important operating variables. The viscosity ranges from 15 to 50 centistokes, depending upon the type of atomizing equipment (i.e., steam, air, mechanical, etc.). Thus, the viscosity of the fuel itself is not critical, but rather the capability of the fuel oil heaters to maintain a constant and correct atomizing viscosity.

Although references have been made regarding the ability of the boiler to burn dirty fuel, there are filters in the fuel handling system to minimize clogging of fuel flow control valves and atomizer tips. To minimize fuel filter clogging, overloading, and eventual deposit build-up on fuel nozzle tips, the fuel must meet an incompatibility requirement. Using the test outlined on Table 29, an incompatibility sediment value of less than 1% is usually sufficient for boiler application.

As the boiler manufacturers increase steam temperature to increase cycle efficiency, ash corrosion of metallic parts (e.g., tubes, tube hangers, etc.) may become a problem.[78] Combustion of fuels containing significant amounts of porphyrins and other metallic salts can result in the formation of low melting point vanadium pentoxide (V_2O_5), vanadates such as $Na_2O \cdot 6V_2O_5$, or alkali sulfates which are corrosive to metals at combustion temperature. Currently, the corrosion is controlled by (a) limiting steam temperature, (b) reducing excess air to minimize oxide formation, (c) adding refractory coating, (d) use of high chromium alloys, and (e) the addition of alkali metals (e.g., magnesium oxides or carbonates) to the fuel to combine with the vanadates and increase their melting points.

Alternate Fuels/Steam Propulsion. Considering the above performance requirements, liquid fuels from alternate sources must still meet an incompatible sediment value. The viscosity can vary, but, as with diesel fuel, the pour point should not exceed 35-40°C unless special precautions are taken to provide adequate heating. Since many of the coal liquids retain some of the trace elements from the coal,[79] studies will be necessary to determine any new high-temperature corrosion problems. Further, there seem to be no serious problems regarding the efficient combustion of coal liquids. McCann and his coworkers reported successful results using SRC[80] with minor boiler modifications. The U.S. Navy ran tests with COED-derived fuel oil and determined that it was feasible to use synthetically derived oil as marine fuel. In addition a trial with shale oil was conducted aboard an ore carrier on the Great Lakes. No changes were made to either the boiler or burners relative to normal Bunker C firing. During the test, no operational difficulties or detrimental effects to combustion were noted.

Studies are also underway on coal/oil slurries as well as pulverized coal and fluidized bed combustion in steamships. The coal/oil slurry work[81] indicates that stable powdered coal/oil slurries can be prepared and successfully fired up to 75% of maximum rating in an existing gas-designed industrial boiler. Longer-term tests will be necessary to fully confirm this application. Studies by Hodgkin[82] indicate that fluidized bed combustors could be adopted for marine service and burn liquid, solid, or gaseous fuel.

5. Conclusion

In today's marine fuel market there are variations in commercial fuel properties. However, for the most part, these fuels perform satisfactorily, since they do meet various key performance requirements for the respective propulsion system. On this basis and analyzing the available data, it appears that many liquid fuels from alternate sources could also be satisfactory if they meet these same requirements. Coal and coal/oil slurries might also be ultimately used provided handling, corrosion, deposits, and wear problems are solved.

This comparison is based on technical performance only. For the ultimate future acceptance of alternate fuels for marine bunkering application, the economic development of these alternate fuels, together with the cost of equipment modification and maintenance, must also be considered.

B.   Railroad Utilization of Alternate Fuels

Conan P. Furber
Association of American Railroads

The railroad industry today, as throughout its history, continues to have a great interest in alternate fuels. Faced with the dual problem of steadily increasing diesel fuel prices and the possibility of insufficient quantities of specification fuel being available in the future, the industry is investigating new sources of energy as well as continuing to develop new means for conserving diesel fuel. In this paper a proposal to establish a railroad alternate fuel research program is discussed.

The term alternate source of energy or alternate fuel has had many meanings to the railroad industry. Beginning in the early 1600's, the horse was the standard means of propulsion for the ancestor of today's railroads, the tramway. The standard fuel during that time was hay and oats. With the development by Robert Stevenson in 1814 of the first successful steam locomotive, the standard fuel shifted from oats to wood. The age of steam brought with it new meanings to the term alternate fuels. Fuel for steam locomotives varied from company to company and also by era, thus making wood, coal, and petroleum products all alternate fuels at one time or another. Steam power also ended man's dependence on wind, water currents, and animals as the only source of motive power.

There were many attempts to provide alternate sources of power for use by the railroads: sails were used; by creating a partial vacuum in a conduit, air pressure was used to propel the atmospheric railroad in England; and stationary powerplants (driven by steam or water power) pulled railcars through use of endless cables. By the end of the 19th century electrical power became technically and economically feasible as an alternate means for propelling railcars.

Electrification of railroads in the United States reached its zenith in the late 1930's in terms of track miles and, although not interchangeable, it would have to be considered as an alternate to steam power rather than the standard source. However, the diesel-electric locomotive, first introduced in 1925, rapidly became the standard for motive power for the railroad industry and all other fuels then became alternates to diesel fuel. In 1977, 150 years after the original 13 miles of track was opened to traffic by the B&O Railroad, the Class I railroads were operating a fleet of 27,573 diesel-electric locomotives; seven steam locomotives; and 216 electric units over a system containing over 325,000 miles of track. By comparison, in 1929 the railroads operated only 22 diesel-electric locomotives compared to 56,936 steam locomotives and 601 electric units. Yet in 1977, with less than one half the number of locomotives, the railroads handled 25% more gross ton-miles than were handled in 1929, due almost entirely to the superiority of the modern locomotive.

Railroads are the most energy-efficient mode of land transportation. They provide a 3:1 to 5:1 advantage in energy savings over trucks and consume only about 3% of the total direct transportation energy. For every gallon of fuel consumed, the railroads produce over

200 ton-miles of freight compared to approximately 70 ton-miles per gallon of fuel consumed by regulated common carrier trucks. Although the railroads are energy efficient and use only a very small percentage of the total transportation energy, because of the enormous size of the industry, in 1976 alone an estimated 3922 million gallons of diesel fuel were consumed at a total cost to the industry exceeding $1.2 billion. This fuel was purchased at an average cost of 31.64 cents per gallon. By comparison, the average cost of diesel fuel from 1953 to 1972 ranged between 8.84 and 10.97 cents per gallon.

With 99% of the railroads' motive power fleet being diesel-electric, the steadily increasing cost of diesel fuel and the uncertainties concerning availability of acceptable diesel fuel are problems of great concern to the railroad industry. The railroad industry is developing, in conjunction with federal agencies, a long-term cooperative research program directed at the utilization of alternate fuels and the conservation of diesel fuel.

The overall proposed energy research program, as presently conceived, consists of three major parts: short-term, medium-term, and long-term.

1. Short-Term

Short-term projects are defined as those which could be implemented immediately and which would require little or no modification to the locomotive or fuel delivery system. Examples of such projects include: fuel additives, fuel saver systems, and fuel metering.

2. Medium-Term

Medium-term projects would include programs requiring moderate retrofitting of locomotives and studies of potential alternate fuels not currently available in commercial quantities. Such studies might include: off-specification fuel limits; coal-based fuels; dual fuels; and monitoring instrumentation.

3. Long-Term

Long-term projects would include studies of fuels requiring new delivery systems, new engine designs, and all other studies which would require considerable investment of capital.

The proposed alternate fuel research program was developed with consideration for the following constraints:

(a) Final testing of any alternate fuel, fuel additive, fuel-saving device, or fuel conservation measure must be conducted in the field under actual operating conditions.

(b) Because of the present severe shortage of locomotives, additional consideration must be given to any testing program which requires removal of locomotives from service or increases the risk of damaging a service locomotive.

(c) Fuel consumption under field operating conditions cannot, in general, be measured within about 5% accuracy.

Influenced by these constraints, the concept of establishing a stationary test bed locomotive engine testing facility gradually evolved. A facility of this type would reduce the downtime and risk to service locomotives involved in the research program. It would provide the capability of testing, under laboratory conditions, concepts which might produce only small, but important, percentage reductions in fuel savings. For example, a 1% reduction in fuel consumption would save the industry about $12 million a year. Final testing must still be accomplished in the field under actual operating conditions; however, through judicious screening, the number of tests and risk associated with each test would be greatly reduced.

As presently conceived, the locomotive engine test facility would be equipped with two-cylinder test engines and a minimum of two full-sized multiple-cylinder locomotive diesel engines. Preliminary testing of alternate fuels, fuel additives, etc., would be undertaken on the two-cylinder test engines. If the preliminary testing indicated that further testing was warranted, it would be undertaken on the full-sized stationary test bed engines. Final testing, where warranted, would be accomplished on service locomotives in the field under actual operating conditions.

Although the locomotive diesel engine test facility is not yet a reality, a tentative research program has been outlined. The first task in the program will be an evaluation of off-specification diesel fuel. With the exception of the Canadian National Railways, which is burning semirefined crude oil, most railroads buy only standard No. 2 diesel fuel in accord with rigid specifications. This task will attempt to identify problems and define fuel limits for the full range of power and operating conditions. In addition, engine wear, lubricant degradation, and engine or fuel system modifications will be addressed. The results of this project could enable the railroads to reduce their fuel costs and dependency on specification fuel.

A separate task is proposed for the evaluation of nondiesel fuels, most of which would require engine combustion system modification. Included in this group would be the alcohols and the many diverse liquid fuels produced from coal, tar sands, and shale oils. Preliminary evaluation would be accomplished through the use of the two-cylinder test engines followed, when warranted, by testing on the full-sized diesel engines. Engine performance, fuel consumption, durability, maintainability, and the effects on engine lubricants will be evaluated.

Also to be included within the research program will be an evaluation of fuel additives. There are a large number of different types of additives currently available, some of which may prove to be beneficial either in the area of engine fuel economy or exhaust emissions. Effective evaluation of fuel additives has been a problem to the railroads because of the degree of accuracy generally obtainable under field conditions. Because of the possibility of a plus or minus 5% error in the field, evaluation under laboratory conditions appears to be the only reasonable means of judging the true value of additives. Again, the two-cylinder test engines would be used for screening purposes followed by testing on the full-sized test engines.

A test facility capable of undertaking an alternate fuels testing program of the magnitude necessary to meet the needs of the railroad industry represents a huge commitment of capital and operating funds. It is doubtful that the railroads, considering their present financial status, can adequately fund such a program. This facility and alternate fuels program can become a reality only as a cooperative program involving the railroads, governmental agencies, suppliers, and private research institutes. Benefits to be derived from the program include reducing dependency on petroleum products; conservation of diesel fuel; and reduction in railroad operating costs and freight transportation costs.

4. Conclusions

Throughout their history the railroads have adapted to many alternate sources of energy. Today, they are continuing to search for alternates to specification diesel fuel. Faced with the dual problem of steadily increasing fuel costs and uncertainties regarding the future availability of fuel oil, the railroad industry is considering establishing an alternate fuels locomotive engine test facility. Through the proposed research program to be undertaken at this facility the possibility exists that railroads would be able to further conserve diesel fuel and to investigate the potential of alternate fuels.

C. Alternative Fuels in Aviation

John G. Borger
Pan American World Airways, Inc.

1. A Status Report

Figure 32 shows that fuel prices, after the tremendous surge in 1973-74, having leveled off for about two years now, have again resumed a steady increase. This increase, however, seems to parallel the rate of inflation, approximately 16% over the last two years. Please note that the figure shows Pan Am fuel costs for B747 operations only; total Pan Am costs are slightly higher, and domestic U.S. airlines show an appreciably lower figure, although the domestic recent rate of increase has been somewhat greater.

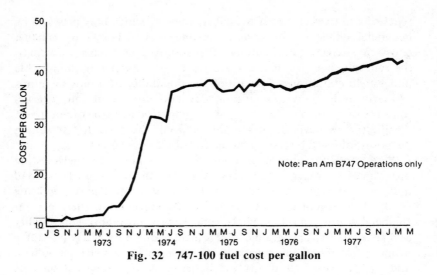

Fig. 32　747-100 fuel cost per gallon

Reference to Figure 33 shows the impact of fuel costs on the direct operating costs of the B747. The fuel portion has risen from about 20% to just under 44%. This fraction is probably higher for other airplanes, for the B747 may be the most fuel-efficient transport. For those not familiar with airline accounting procedures, it should be noted that direct operating costs do not include all operating costs; indirect costs frequently equal or exceed the direct cost.

For the past three years, the fuel fraction of direct operating costs has remained relatively constant. This does not imply that the direct costs themselves have remained constant, for reference to Figure 34 shows a steady increase in the total fuel cost per hour of operation, in spite of cyclic variations in fuel consumption.

These variations are attributable mostly to variations in payload carried. Fuel consumption in the third quarter is consistently higher than that for the first quarter, because payloads are much higher. Fuel burned will also vary with length of flight, but these charts are for one airline, and schedule variations are relatively minor. Some slight improvement has been due to operating slightly longer average stage lengths.

The variation of fuel burned with payload is further developed in Figure 35. This shows the increase in fuel consumption with increases in dry tank weight (operating weight empty plus payload). Dry tank weights for the present summer may exceed 430,000 lb, for the seating capacity of the airplanes has been increased from 375 to 405, and loads are currently running quite high. It is interesting to note the fuel consumption improvement that took place between 1975-76 and 1976-77: about 2%. This is mainly attributable to improvements in

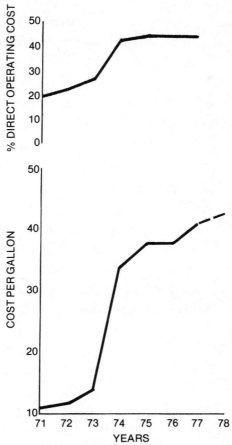

Fig. 33   747-100 annual fuel cost summary

operating techniques, some clean-up, and some improvement in engine efficiency due to a modification program.

Figure 36 presents recent history of fuel consumption and improvements achieved. Although the actual gallons per hour have increased, the dry tank weights have increased sufficiently so that there is actually a net improvement in basic fuel consumption. The dip in the improvement curve is more difficult to explain, but is believed attributable to engine fuel consumption deterioration.

This brings up the subject of the most serious of our current problems: engine performance deterioration. The modern high bypass turbofan engine with its high turbine inlet temperatures and high compression seems particularly susceptible to progressive increases in fuel consumption and turbine temperatures; the latter seem to induce

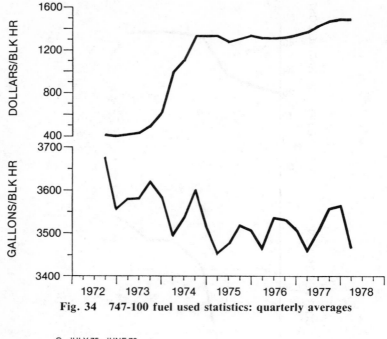

Fig. 34   747-100 fuel used statistics: quarterly averages

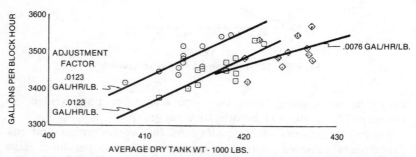

Fig. 35   747-100 average dry tank weight versus gallons per block hour

still further temperature increases. Fuel consumption deteriorations as high as 6½-7% have been reported, and all the high by-pass turbofans are susceptible. Pan Am's experience with the JT9D-7 indicates a maximum of 3½-4%. Some of this deterioration can be traced to erosion of airfoils within the engine, some to increased rotor tip clearances due to case flexibility, and some to increased service clearances after engine repair; but these do not provide all of the

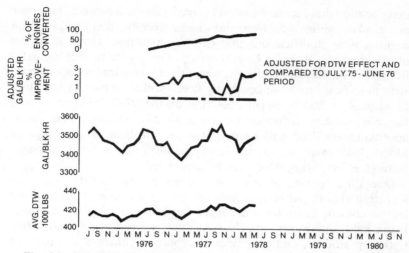

Fig. 36   747-100 reduced fuel consumption performance improvement

answers. NASA has sponsored a program on engine diagnostics, which has yielded some answers. Pan Am has undertaken with Pratt & Whitney a major engine refurbishment program intended to improve engine performance after repair or overhaul. To date, the results are rather promising, with an indicated improvement of 1-1½%, and expectation of somewhat better improvement in the future. This program, initially at least, results in increased material and labor costs; if, as intended, it also results in reduced engine shop visit rates, it will be economically justifiable.

There is no question but that a major effort by the engine manufacturers and operators will be required to minimize this performance deterioration. Unfortunately, this will all be a rear guard action, for the objective is to maintain the original performance of the engines, not to improve it.

Other conservation measures have been put into effect, and the effort is continuing. The most rewarding way to increase fuel efficiency is to increase the capacity of the airplane; that is, put more bodies in the seats or more cargo in the cargo compartments. Virtually all airlines have increased seating capacities in their airplanes, either by reducing First Class seats and devoting more space to Economy seats, or by increasing density of Economy seating. However, while this may improve fuel efficiency in terms of seat (or ton) miles per gallon, it does not necessarily decrease total gallons consumed. On the contrary, with the increase in loads resulting from the marked decrease in fares, total airline consumption undoubtedly will increase.

There is still room for improvement with current equipment. Not only can we endeavor to alleviate the effects of performance deterioration and increase payload capacity in the airplanes, but there are airplane drag reduction and engine specific fuel consumption improvement modifications that can be retrofitted. Often these are quite expensive, and may be cost effective only with further increases in fuel prices; almost invariably, much time is required to incorporate such modifications in an operating fleet. And we must pay attention to weights. A 200-lb weight reduction on the B747 could result in a fuel consumption reduction on the order of 1.75 gallons per hour, approximately 7500 gallons per year per airplane or a savings of $3400. Like many of us, airplanes usually grow heavier with age; through effort, aging effects can be minimized.

Operating techniques are also susceptible to further fuel conservation efforts. Although there has been some improvement in air traffic control handling procedures, we still see cases of airplanes being held unduly long in clear weather, operations at lower than optimum altitude, and long lines at departure runways. Airlines are also testing new flight management systems intended to assist the pilots, or actually control the altitude, airspeed, and throttle settings to maintain more nearly optimum conditions, without the plus and minus variations that bring on additional fuel consumption. Such devices should reduce overall fuel consumption by 1½-5%, depending on the precision of the current manual techniques.

Fuel efficiences of various Pan Am airplanes are shown on Figures 37 and 38. Figure 37 is plotted in terms of payload tons per 1000 lb of fuel burned; obviously this decreases markedly as range increases. The more modern airplanes show continuous improvement. Figure 38 is in

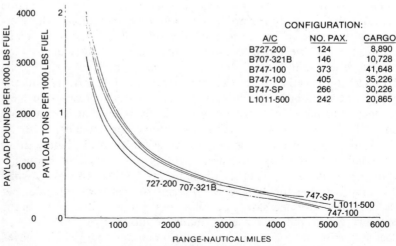

Fig. 37   **Payload per pound of fuel**

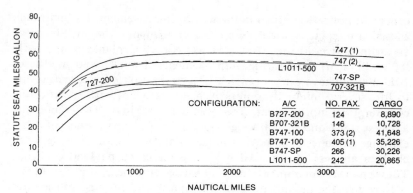

Fig. 38   Comparison of statute seat miles per gallon

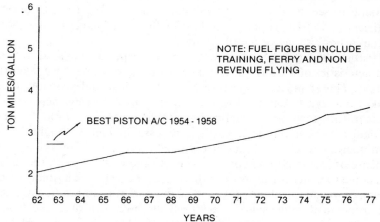

Fig. 39   Fleet ton miles per gallon

terms of seat miles per gallon. This shows the effect of a change in seating capacity on the B747 from 375 to 405. Drag, weight, and specific fuel consumption improvements incorporated in the L-1011-500 show up in its efficiency curve, approximating that of the larger B747 (although the latter curve is for a B747 with lower seating density). Note also that the B747-SP curve is lower by comparison because of the high percentage of First Class seats (44/222) in its long-range arrangement. A more comparable arrangement would be 26/254 = 280.

The overall record of fuel efficiency improvements over the years is shown in Figure 39. The jet airplanes passed the best of the piston airplanes some time ago.

The next step is to incorporate similar and further improvements in new airplanes. The airplanes being built under the current rush of orders do not reflect much in the way of improvement, but inasmuch as they are replacing older, less efficient airplanes, they will improve an airline's overall fuel efficiency. Derivative airplanes of the early '80's will incorporate further improvements. The DC-9-80 and the L-1011-500 will have wings of increased aspect ratio to decrease drag due to lift, several minor drag improvements, engines with specific fuel consumption improvements (the refanned JT8D in the DC9 represents a bigger step), and several weight-saving items. Later models of other existing airplanes will have similar changes, although wing span changes are less certain. All will have increased payload capacity. These changes can improve fuel efficiencies anywhere from 5 to 15%.

New airplane designs now on the boards will provide still greater improvements. Here there is an opportunity to optimize the design as a whole, rather than incorporate modifications by bits and pieces. It is difficult to see, however, that benefits of the NASA Aircraft Energy Efficiency (ACEE) program will show up in new airplanes before 1985 at the earliest; it may very well be later.

Historically, the better way to insure superior payload range characteristics of a transport airplane has been to increase the takeoff weight. This requires an increase in empty weight and probably fuel tank capacity. Usually these are accompanied by an increase in takeoff thrust so as to maintain suitable takeoff and climb performance, which means a further increase in empty weight. The combination results in reduced cruise altitude and is always at the expense of increased fuel consumption. In the writer's view, the time has come to endeavor to unwind this cycle. If careful attention is paid to empty weight, airplane drag, and engine fuel consumption, equal or better payload range can be achieved at lower or equal fuel consumption. As an example, preliminary studies have shown that a B747 designed to these principles can have the same payload range as earlier B747's but with fuel efficiency 10-15% better.

To this point, we have concentrated on means of improving fuel efficiency or, in other words, decreasing fuel consumption per unit of payload carried. The other means we can resort to is use of some other type of fuel. The most extreme of these is liquid hydrogen, then methane. Perhaps something will come along in the form of a fuel that is liquid in the normal temperature range, but produces more Btu, and therefore would exhibit lower consumption. The current specification for jet fuel could be broadened, thus permitting more crudes to qualify. And finally, jet fuels could be obtained synthetically from sources such as coal or shale.

Let's discuss these in turn with regard to their possible use on transport airplanes.

2. Liquid Hydrogen

This is a cryogenic fuel, which must be maintained at very low temperatures to remain liquid. Btu content per pound is highest of any known combustion fluid. The turbine engine is most easily adapted to the fuel; only relatively simple changes must be made to adapt the engine fuel control to the lower density of the hydrogen. Use of hydrogen would require completely new airplanes, with extremely large insulated storage tanks. The airplane must virtually be designed around the tanks. But by far the most difficult obstacle to overcome is the logistical problem of getting the fuel aboard the airplane. Assuming that gaseous hydrogen would be produced at some plant within 100 miles of a major airport, the product would have to be transported, preferably by pipeline, to the airport where a liquefication plant would be built. This liquefication plant, for safety reasons, should be located in a remote area of the airport, with vacuum shrouded distribution pipelines to the various gates. Presumably, refueling could not be conducted while cargo or galley loading equipment is functioning, because of possible explosion hazard. In a study conducted for NASA, it was estimated that airport facilities (not including the gaseous production plant) at a typical airport (San Francisco was used as an example) would cost $340 million in 1975 dollars. Delivered hydrogen would cost almost four times the amount per Btu as jet fuel. It is difficult to envisage the practicability of hydrogen as a fuel for air transportation.

3. Methane

Methane also has a very low boiling point, and in some respects is similar to hydrogen, but with lower energy content, and somewhat more density. Many of the practical problems typical of the use of hydrogen will accompany the use of methane, but perhaps not to as severe a degree. Because of the higher temperature and density of liquid methane, handling and other costs may not be as great as those for hydrogen. It is therefore suggested that any further development of hydrogen as aviation fuel be conducted in parallel with similar consideration of methane.

4. Broadening of Turbine Fuel Specifications

It has been suggested that a greater fraction of petroleum-based fuel would be made available if the specifications for turbine fuel were broadened. This was discussed in detail at a workshop held at NASA Lewis Research Center, June 7-9, 1977. As a result, a specification for an experimental reference broad-specification fuel was drawn up, and NASA will direct future research effort at developing compatibility of turbine engines with such a fuel. The most important changes from current specifications are:

(a) Increased aromatics (decreased hydrogen) content.
(b) Decreased flashpoint (increased volatility).
(c) Increased freeze point.
(d) Lower breakpoint temperature for thermal stability.

The research program will be fully justified if it shows that such broadened specifications will truly make more fuel available without introducing unsolvable operating problems or exposing operations to new safety problems.

Broadening of the fuel specification very probably will render it more difficult to meet proposed new emissions requirements. Some tolerance on the part of EPA may be in order.

5. Increased Aromatics

Aromatic content of fuels is roughly inversely proportional to the hydrogen content. As U. S. and certain other crudes have become exhausted, other crudes becoming available have higher basic aromatic content, and many refineries have difficulty producing fuel to meet the old maximum of 20%. Two years ago an increase to 22% as required was approved on a trial basis. A long-range operator's concern revolves around the facts that decreased hydrogen content means less Btu per pound (although more Btu per gallon) and the fuel burns with a redder flame, thereby increasing temperature of the metal of the various hot parts, which are by far the most expensive parts in a jet engine. If the NASA program can develop means for the engines to tolerate such fuels without major effect on component life or material costs, it will be a worthwhile effort. Some concern is expressed, however, if more complex dual nozzle combustors are required for this purpose; these are much more expensive, will undoubtedly introduce new maintenance problems, and may introduce safety considerations.

As mentioned above, the high by-pass turbofan already runs at quite high temperatures and is quite sensitive to temperature increases. Some of the research effort to develop engines exhibiting lower fuel consumption is based on further temperature increases.

6. Decreased Flashpoint

Current flashpoint requirement is 100°F minimum. Proposals have this going as low as 80°F. The major difference between aviation kerosene and JP-4 wide-cut fuel is in flashpoint, with the latter set at 10°-15°F minimum. Most airlines have abandoned the use of JP-4, except at airports (particularly military) where kerosene is not available. Now the military has started to change over to a higher flashpoint fuel, JP-8. While 80°F is still well above this value, it is below ambient temperatures at many airports during the summertime; thus an increased hazard would probably exist in the event of a fuel spill. In any event, this endeavor appears to be counter to the effort to increase freeze point.

7. Increased Freeze Point

Current specifications call for − 40°C for Jet A and − 50°C for Jet A-1, which is primarily used in international flights. Virtually all domestic U. S. operations are based on Jet A, as are about half of Pan Am's operations, all from U. S. bases. There is good reason to believe the 50°C requirement could be raised somewhat, based on Pan Am's experience on New York/Tokyo and San Francisco/Hong Kong operations. Higher levels, such as the suggested − 20°F for the NASA reference fuel, will undoubtedly require some type of fuel tank heating in the airplane in order to maintain the liquidity of the fuel. It is pointed out that this temperature is above nighttime winter temperatures at many U. S. airports; therefore, the oil companies may have to provide lower freeze point fuels during the winter, just as they provide lower freeze point heating fuels in Canada and northern states during the winter.

As a personal opinion, the writer feels this is a direction in which the industry should move. Modifications to aircraft to insure maintenance of the liquidity of fuels should not be too difficult to incorporate, either in new airplanes or as retrofit; there is plenty of waste heat available around an airplane.

8. Thermal Stability

This has not been a current problem, except possibly in the Concorde. However, in the drive towards higher operating temperatures of jet engines in the effort to realize lower fuel consumption, thermal stability may be of greater importance.

9. Synthetic Fuels

Aviation fuels have been made on an experimental basis from coal and from shale. These are really fuels from alternative sources, for although specification characteristics may vary somewhat from those of present Jet A, the endeavor is to produce fuels compatible with the present engines. The primary problems with both are those of producibility and ultimate cost. One other problem with production of fuel from shale is an environmental question: what to do with the waste shale from which most of the hydrocarbons have been extracted?

Jet fuel from shale apparently will be more suitable for turbine engines. There may be some problem with nitrogen and possibly sulfur, but again, the big problem is getting the process into production. Presumably this will occur when the price of natural petroleum rises to a point where the synthetic product will be competitive.

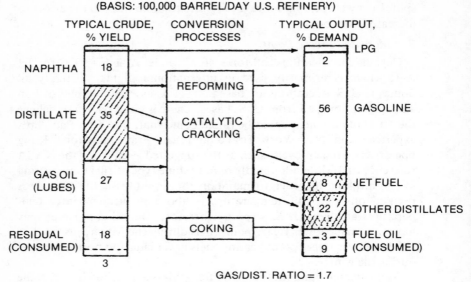

Fig. 40   **Product yield versus demand**

Jet fuel from coal shows indications of high aromatic content, and therefore will be subject to all of the questions that apply to other high aromatic content fuels. Sulfur content also may present a problem.

10.   The Refinery Process

Without posing in any sense as being very knowledgeable regarding the process of oil refining, I have observed that since the turn of the century the refinery process was directed at producing more and more motor gasolines. With the growth of home heating with oil, more and more use of diesel engines, and the minor role played by aviation fuels, more attention is currently being paid to the so-called middle distillates. This would indicate a switch in the direction of refining to produce a greater percentage of such products. One concept is shown on Figures 40 and 41.

While the portion of middle distillates required by aviation is a relatively small percentage of the total, the specification requirements may be more stringent, except possibly for the diesel fuels. This means that we in air transportation will be competing for fuel with other users, whereas 20 years ago, in the U. S. at least, fuel producers were happy to unload kerosene. This may imply a further relative increase in prices.

By now air transportation has become a fixture in our way of life. Unless we are to revert back to trains or ships, now virtually unused

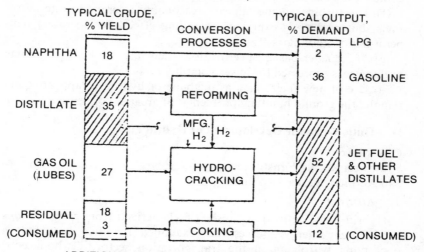

Fig. 41 **Maximizing distillates in future refinery**

for personal transport, some way should be found to support air transportation.

My personal view is that there seems to be no major obstacle to continued use of middle distillate fuel by air transportation; possibly more attention should be paid to the lower volatility end.

The hydrocracking concept gives rise to a further thought. If, instead of making a complete switch to hydrogen fueled airplanes, a way could be found to introduce more hydrogen molecules in liquid hydrocarbon fuel with resultant increase in energy content, conceivably there could be a stretchout in petroleum supply.

11. Conclusions

(a) The airlines have made progress in conserving fuel; this will continue but to a lesser degree. Government contributions to fuel conservation have helped, but could be better.

(b) Derivative transport airplanes will appear in the early 1980's with further small improvements in fuel efficiency.

(c) Derivative designs with greater degree of modification or new designs incorporating more radical changes could appear in four to five years. These could exhibit at least 10% improvement in fuel efficiency, possibly more than 20%.

(d) Major fuel efficiency improvements resulting from the NASA ACEE program probably will not appear before the late '80's.

(e) To date, no substitute for liquid hydrocarbon fuels has been demonstrated to be suitable for transport airplanes.

(f) The turbine engine appears reasonably tolerant to alternative fuels, but caution is expressed regarding effects on operating temperatures of "hot parts."

(g) If broader-based specification fuels are to be used, some preference is expressed for lower volatility fuels.

(h) Use of new fuels, such as methane or hydrogen, appears quite remote; the ground handling problem is of greatest concern.

D. Outlook for the Development of Hybrid Fuels

C.A. Moses
Southwest Research Institute
San Antonio, Texas

1. Abstract

Hybrid fuels, defined as finished fuels derived from combinations of different energy sources, are examined for their potential in conserving petroleum in the transportation sector. Carbonaceous slurries and diesel fuel blends with water and/or alcohols offer the greatest potential. Problems appear solvable in the near future.

2. Introduction

Worldwide petroleum reserves are a nonreplenishable commodity. The rate of depletion depends primarily on the economic growth rates of all the countries of the world and the production rates allowed by the oil producing nations.

Currently the oil producing countries are meeting the world's demand. Sometime in the future, the OPEC will have to decide whether to limit production and extend their supplies or to continue to meet the demand as long as possible. Regardless of the choice, at some point in time we, and the world in general, will be faced with supplementing our petroleum production with other energy sources to meet the demands of economic growth.

The transportation sector consumes large volumes of fuels and is a national target for the conservation of petroleum energy. Petroleum fuels can be conserved by replacing them with hydrocarbons from other sources, e.g., oil shale and coal, or by blending them with non-hydrocarbon as well as hydrocarbon components from nonpetroleum sources. This is likely to result in the use of unorthodox fuels, and some hardware changes may be necessary in order to accommodate fuels in their most economical and abundant compositions.

For the purposes of this paper, finished fuels derived from combinations of different energy sources are called *hybrid fuels*. Such fuels can be slurries, emulsions, and homogeneous solutions, but can also include "dual fuel" concepts where the two fuels are not premixed but introduced into the combustion chamber by separate delivery systems. The supplementary component would necessarily come from a renewable source, e.g., alcohol, or an extremely abundant one, e.g., coal.

The use of hybrid fuels to extend the supplies of petroleum-derived fuels is not only a viable alternative in the short- to midterm until hydrocarbons from oil shale and/or coal become available in significant quantities, but will extend those supplies as well. Some hybrid fuel concepts could also permit the utilization of lower quality fuels, thus reducing the refining costs of fuels from syncrudes; this can be accomplished by tailoring the supplement(s) to achieve certain combustion characteristics such as lean combustion, cleaner burning, or reduced emissions.

Examples of different kinds of hybrid fuels are given below:

(a) Single Fuel

(i) Slurries (solid-in-liquid): powdered carbon (or coal) in fuel oil; wood pulp in fuel oil.

(ii) Emulsions (liquid-in-liquid): water-in-fuel; alcohol-in-fuel.

(iii) Slurried emulsions (liquid/solid/liquid): fuel/carbon/water; fuel/carbon/alcohol.

(iv) Solutions (liquid/liquid): alcohol fuel; X/fuel (X = some liquid).

(b) Dual Fuel

(i) Gas and liquid: hydrogen and gasoline; hydrogen and diesel fuel; natural gas and diesel fuel.

(ii) Liquid and liquid: alcohol and diesel fuel.

3. Hybrid Fuel Experience

Examples of the use of hybrid fuels are presented to illustrate the potentials and problems of this fuel concept.

Single Fuels. (a) Slurries. This class of hybrid fuels is best represented by a slurry of pulverized carbon in fuel oil. The term "carbon" should be taken in a broad sense to mean predominantly carbon and could be represented by a number of solid compositions, e.g., coal, charcoal, wood pulp, or lignin. The first proposal to use coal-oil mixtures appears to have been made in 1879, so the concept of hybrid fuels is not new.[83] In fact, some of the early justification was to reduce oil imports (to England) and increase the market for domestic coal. These early experiences were generally for ship boilers and not for IC engines.[84] A fairly complete review of early applications for the diesel engine is provided by Hanse[85]; most of this work was applied to large-bore, slow-speed diesels. Problems were those of grinding the coal, stabilizing the slurry, and excessive wear of fuel pumps and injection equipment. Two recent studies by Marshall et al.[86] and the U.S. Army Fuels and Lubricants Research Laboratory[87] will serve as examples of recent efforts oriented towards the small-bore diesel for transportation. Marshall et al. at VPI operated a 1360 cc diesel engine on a slurry of 15% solvent-refined coal in jet fuel. The coal was pulverized to a nominal size of 2 micrometers. This slurry would flow through the injection nozzle and form acceptable spray patterns. Test results showed fuel consumption to be generally better than pure Jet A but not quite as good as DF2; the differences were not great. The Army has been investigating slurries of carbon black and diesel fuel for the purposes of increasing the energy density of the fuel, thus effectively extending the range of a vehicle. Engine tests with slurries of up to 50% carbon black have shown that as much as 90% of the heating value of the carbon can be recovered. Both programs have identified injector wear or plugging as problems but have shown that pulverized carbon can be efficiently burned in a diesel engine and potentially replace a sizable fraction of the diesel fuel consumption. Further work is necessary to solve the fuel handling problems and optimize the combustion processes. Slurry stability is still a problem, but recent work by Ashland Oil and the Army shows promise of solving this with submicron sized carbon particles and surfactants.[87,88] Other slurries which could be considered are wood pulp in its various forms or stages.

(b) Emulsions. There has been a renewed interest in the combustion of emulsions in the last 5 years, much of which has been oriented toward continuous combustion systems such as boilers and turbine engines. Dryer provides an excellent discussion of the combustion of emulsions.[89] Most of the work has been with water-in-fuel emulsions where the water does not contribute any energy but has been shown to modify the combustion process, for example, to reduce exhaust smoke and NO_x in diesel and turbine engines,[89, 90] and to improve the combustion efficiency of fuels more viscous than the system was designed for.[91] The use of water-in-fuel emulsions for diesels has been sparse and conflicting. Studies at the Cummins Engine Company found that the ignition delay was so markedly increased that no benefits could be obtained except for a reduction in smoke.[92] However, Kahn at CAV Ltd. was able to improve brake specific fuel consumption in addition to reducing smoke and NO_x by adjusting the injection timing.[89] Storment at Southwest Research Institute found 2-5% improvements in fuel consumption with 10, 15, and 20% water volume, accompanied by substantial reduction in NO_x and smoke but increases in CO and UBH.[93] Both the Cummins and SwRI work were with unstabilized emulsions. The Army has conducted work on both ordinary and microemulsions of water-in-fuel. In both cases, the decrease in power was less than the amount of water added, also indicating a stretching of the hydrocarbon fuel supply.[87] Recently, some work has been done on alcohol in diesel fuel emulsions to avoid the energy penalty of the water. Ontario Research Foundation has conducted some cursory tests of emulsions of 20% methanol in diesel fuel in a Duetz engine. They found a 5-6% reduction in peak power which can be translated to an increase of 5-10% in thermal efficiency when heat content is considered. The Army has found similar results with no effects on ignition delay, complete combustion of the alcohol, and reductions in smoke; there is also some preliminary indication that the thermal-cycle efficiency has been improved.[87] Some work on water-in-gasoline and alcohol/water/gasoline emulsions was conducted in the early 1970's by the U.S. Postal Service Maintenance Technical Center and by the Army. The water/gasoline program was suspended because of phase separation problems; the alcohol/water/gasoline programs both resulted in significant increases in fuel consumption, power losses, and higher fuel costs.[94]

(c) Slurried Emulsions. The Army has also conducted brief experiments[87] with three-phase slurried emulsions of fuel, carbon, and water, their purpose being to find a fuel that is fire resistant as well as energetic. They reported good energy performance, but continuous stirring of the fuel was required to prevent settling of the carbonaceous phase. More exploratory work appears to be required in this area before an assessment can be made.

(d) Solutions. Mixtures of alcohol and gasoline have been studied for some time under a number of federally funded programs. Recently, several states in the Midwest have eased the way for such fuels to enter the commercial market. Generally speaking, blends of 10-15% methanol in gasoline present no serious performance problems and can be used in current automobiles without hardware changes. There appear to be two major problems: one is a phase separation which takes place if water is absorbed by the alcohol; no solution has yet been found. The other problem is compatibility with the engine lubricant; this will probably be solved with a new additive package. A concept which has apparently not been investigated is a solution of alcohol in light or middle distillate such as kerosene or diesel fuel containing sufficiently high concentrations of aromatics in which the alcohol is soluble. (This is different from the alcohol/fuel emulsions already discussed.) An interesting possibility is the distillates made from coal which contain very high aromatics and currently must be hydrogenated at great expense to produce acceptable diesel and aviation fuels. Blending with methanol (also produced from coal) would increase the hydrogen content to acceptable levels for aviation use and also provide vapor pressure for use in spark ignition engines. (Aromatics are commonly used to raise the octane level of unleaded gasolines but perform poorly in compression ignition engines.) This would not only serve to extend the hydrocarbon fuel but possibly be cheaper than the refining processes.

Dual Fuels. Many compression-ignition engines operate on the dual-fuel principle. For these engines the primary fuel is usually gaseous at atmospheric pressures and temperatures and is normally inducted with the air. At some point near compression top dead center, a charge of diesel fuel is injected through the conventional diesel fuel system to act as a source of ignition for the compressed fuel-air mixture. This pilot charge usually represents less than 10% of the total fuel. This concept is often used for economic reasons in regions where natural gas, propane, and others cost less per Btu than diesel fuel. Other advantages are less lubricating-oil dilution, cleaner exhaust, and less engine wear. The power output is usually limited by loss of combustion control, which is similar to knock in spark ignition engines and can be controlled by antiknock compounds.[95] Similar problems have been encountered in experimental studies with hydrogen as the primary fuel.[96] Hydrogen has also been investigated by JPL (e.g., Ref. 97) and GM (e.g., Ref. 98) for use in spark-ignition engines but not as the primary fuel; instead advantage was taken of the wide flammability limits of hydrogen to allow combustion at ultraclean conditions to prevent NO_x formation. While NO_x could be significantly reduced, CO and UBH were unaffected or substantially higher because of the cool flame combustion. Overall efficiencies (including an 80% efficient hydrogen generator) were down about 4%. Liquid fuels have also been tried as the primary fuel in dual-fuel CI engines—generally methanol and ethanol. The people at Riccardo carburetted the methanol into the intake air.[99] The pilot charge was required to be quite large to overcome the quenching of the evaporating methanol; inlet heating and amyl nitrate were helpful as ignition improvers but brought the quench and knock limits closer together, making practical operation difficult. Volvo got around this problem by injecting the methanol during the ignition delay period of the pilot charge; efficiency was the same while smoke and exhaust temperatures were lower. Road tests were conducted on as much as 91% methanol by volume (82% by energy) at high speeds and loads. Endurance tests showed anticipated problems with some gaskets in the fuel system and unexpected injector faults caused by a lower damping performance of the methanol when the needle hit the seat—both correctable.[100] Ethanol is a strong candidate in India because it is a largely agricultural country. The results of work done at the Indian Institute of Technology in Madras are similar to the results on methanol[101] with as much as 70-80% of the heat requirement obtained from alcohol over most of the load range.

4. Summary

Most of the categories of hybrid fuels have been tried with varying degrees of success. All of the solid/liquid and liquid/liquid combinations (with the exception of water emulsion) appear to have a significant potential for replacing the hydrocarbon fuel with more plentiful or renewable fuel, thus extending the hydrocarbon resources or permitting the use of lower quality fuels. Only the methanol/gasoline blends for SI engines appear imminently ready for the commercial market. The addition of carbonaceous slurries and alcohols to diesel fuel offers the greatest potential for extending supplies; the problems of application to mobile engines do not in general seem insurmountable and could be solved in the near future, say, before 1985—in time to still be of use in extending petroleum supplies until coal and oil shale products come on stream in significant quantities. Alcohol hybrids also offer good potential for reducing the refining requirements of coal-derived liquid fuels for aviation fuels and spark-ignition engines.

E. Commentary by Symposium Participants

(a) The fact that hydrogen, for example, is four times the price of current Jet A fuel is irrelevant. The only proper comparison of fuel prices is one where we look at one synthetic against another synthetic, and try to determine what their respective costs will be at some future date. As for the cost of the system for supplying liquid hydrogen at the airport, the major portion of that goes toward the liquefaction of the hydrogen. NASA has studied this and arrived at a typical figure, for a large airport, of four to five hundred million dollars—when only the wide-bodied aircraft are considered. This figure is comparable to the cost of current airport modifications (of which, however, not all are technical).

(b) As we go about making a decision on which alternative fuels we should develop, we must look at still another element besides the ones that have been discussed so far: which way we can expect other countries to go. In many cases, that will be determined by what they have to work with. Many countries do not have a lot of coal, as we do, or a lot of oil shale, as we do. They may have to go to hydrogen because they have nothing else. As for the initial lack of hydrogen facilities at very many airports, we should remember that we coped with a similar problem when Jet A fuel was introduced. Until it was available at every airport, planes that used it had to fly only to airports that handled it.

(c) There are still many unanswered questions as to the safety aspects of hydrogen in the event of a plane crash. Its greater volatility suggests a greater likelihood of fire. NASA would like to analyze

accidents that have occurred to determine not only the frequency of
fire, but its effects when it happens: How many people are killed by
the actual fire? How many by smoke? This latter statistic may suggest
a positive side to using hydrogen. The question of safety at the airport
in fueling the plane has been studied by Brewer at Lockheed. He
examined various hydrogen fueling systems, and concluded that the
safest is one in which any vent gas released during the fueling is
recaptured and returned to the plant for reliquefaction and reuse.
Thus, no hydrogen escapes, and fueling can take place simultaneously
with passenger and cargo loading, just as is done now. As for the cost
of liquid hydrogen compared to other synthetics, studies at Lockheed
show that coal-derived liquid hydrogen and coal-derived synthetic
aviation kerosene are competitive, but at this point we know much
more about producing the hydrogen than we do about producing
kerosene. The main obstacles to the use of liquid hydrogen, and they
are considerable, are the need to design completely new airplanes to
handle it and the logistic problems on the ground that would have to
be solved. There is also a feeling by some that the relative safety of
hydrogen may be overstated by its proponents. Liquid hydrogen as an
alternative fuel is an issue that has become rather emotional on both
sides. It would be well to separate out the emotion and remember that
liquid hydrogen is a long-term option, not a short-term one, and that
it cannot be either embraced or dismissed at this point. We need first
to get all the data so that we can refine its problem areas—safety, cost,
logistics, and the rest; and only then can we begin to make any
decisions on its viability.

(d) *Question:* Would not the use of the broad-specification experimental reference fuel for aircraft that NASA is working on be a
step backward in terms of engine maintenance, engine life, and
emissions?

Answer: It is being developed precisely to determine these effects and problems. Once they are understood, we can project the
degree of complexity needed in redesigning or modifying engines.
Then, together, the engine people and the fuel people can arrive at
practical answers in putting together some broader-spec fuel with a
new or modified engine, and still meet all standards for engine
maintenance, life, and emissions.

(e) *Question to the panel:* What are the implications of the lifetimes
of engines in the transition to alternate fuels?

Barry: U.S. flag vessels are changing from steam turbine to
diesel, and the change is expected to accelerate. The reason is that
large-bore, slow-speed diesel engines are more fuel-efficient than
turbines. The manufacturers are building engines that are expected to
last for at least 20 years, and since all sorts of changes in fuel can be

expected over that period, engines are being modified to give them the capability of burning the widest variety of fuels, from liquid to solid.

Borger: Jet engines cannot be characterized as having a "lifetime." By the end of five years of use, every part of the engine but the case and the nameplate has been replaced. This makes it relatively easy to retrofit with improved parts as they are developed and is, in fact, one of the ways in which we have been improving fuel consumption. We therefore have no real problem in adapting engines to fuel changes as they come on stream. The main exception to that would be for cryogenic fuels. These would require a complete redesign of the airplane.

Furber: There are better than 27,500 locomotives in service, some that go back to before 1953. A new one costs between $650,000 and $800,000—so they are built to last for as long as thirty years. It follows that it would take a long time before any engines of new design could make an appreciable dent in that number. The existing engines are extremely efficient and we don't look for too much more there.

F. Workshop Session Discussion (Road and Nonroad Vehicles)

There is general agreement that the changes in fuel and engine utilization will be evolutionary rather than revolutionary. There simply is so much "inertia" in both the engine and fuel systems that revolution can take place only in a particular subsystem. For example, if hydrogen were to be used as an aircraft fuel, the entire plane and fuel system would have to be changed.

A second consensus is that present engines will continue to dominate the field; i.e., a modified spark-ignition engine will provide power for the automobile, and diesel engines will be used for trucks. There would be some penetration of the gas turbine in the heavy-duty-truck field and in the passenger-car field as well as increased use of turbocharging.

Thirdly, it is primarily engineering changes rather than scientific breakthroughs that are needed in order to use most alternate fuels. Techniques for burning alcohol in spark-ignition engines are known. Most liquid alternate fuels might have different aromatic or naphthenic fractions that would require development work in order to be utilized. One possible exception would be direct use of powdered coal in reciprocating engines.

The fourth and most important consensus is that the inevitably large inertia and long delays in the system mandate an immediate start on the implementation of these evolutionary changes. Most of us believe it will take a minimum of 10 to 15 years to *start* commercial production and several times that long to completely switch the system to alternate fuels. Other conclusions of this session have been stated in Section I of this book.

•

Because of the obvious importance of timeliness in the introduction of alternative fuels, a survey was conducted of all of this session's participants, who represented perhaps one of the most knowledgeable groups on this subject ever convened, as an attempt to identify a realistic implementation time scale. Participants were asked to give their best estimates of alternative fuel utilization for the arbitrarily chosen years 1990, 2000, and 2025. For the purposes of the survey, it was assumed that the totals of alternative transportation fuel in use at those times would be 5%, 15%, and 50%, respectively. The participants were to estimate the proportion of those percentages taken up by specific alternative fuels, and to predict the sources of those fuels.

Figures 42, 43, and 44 illustrate the results of the survey. In the charts, the total alternative fuel utilization fraction in each end-use form is subdivided into ten equal portions of the abscissa (horizontal) axis, each representing one tenth of the total percentage-in-use assumed for the particular year. For example, in Figure 42 each abscissa division represents 0.5% of the total usage of an alternative fuel. The sources of the fuels are shown on the ordinate (vertical) axis, and each increment represents a one-person "vote," i.e., a judgment of one individual that that particular fuel source will be utilized in a particular amount (fraction of 5%) for each of the abscissa end-uses. Thus, in Figure 42 (the "5%" chart), at the intersection of "Biomass" and "Alcohol," we see that 5 people thought that in 1990 alcohol from biomass would comprise from 0%-0.5% of the total alternative fuel in use; 10 people thought it would be from 0.5%-1% of the total; 1 person thought it would be from 1%-1.5%; etc. At the intersection of "Coal" and "Alcohol," 1 person thought that alcohol derived from coal would make up between 0% and 0.5%, etc.

Note that the "total" bands of each of the end-use (abscissa) columns do not represent the sum total of all the alternative fuel sources, but only the numbers of participants who believed that the designated fractions of alternative fuels would be used in that application. The "total" bands in each alternative fuel source (ordinate) *do* represent the total use of that source.

The general results of the survey illustrated by Figures 42-44 can be summarized as follows:

(a) When alternative fuels make up 5% of the total transportation fuel supply, the largest portion will come from shale, then coal, then

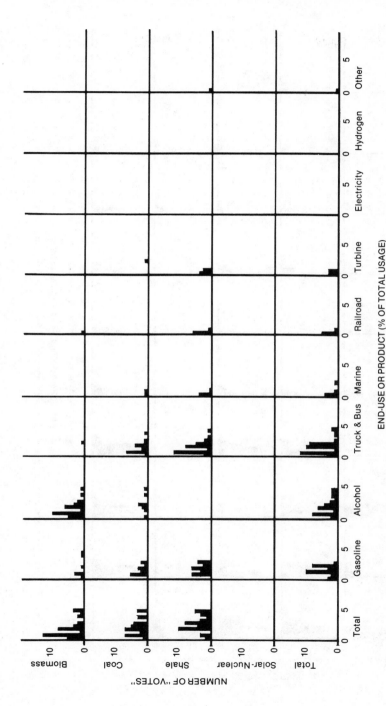

Fig. 42 Energy source and products when alternate fuels are 5% of transportation energy

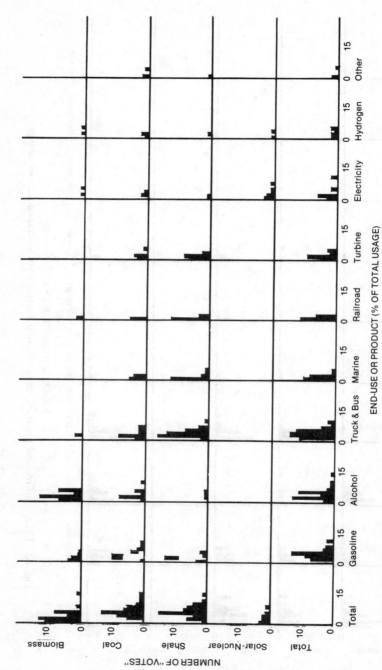

Fig. 43   Energy source and products when alternate fuels are 15% of transportation energy

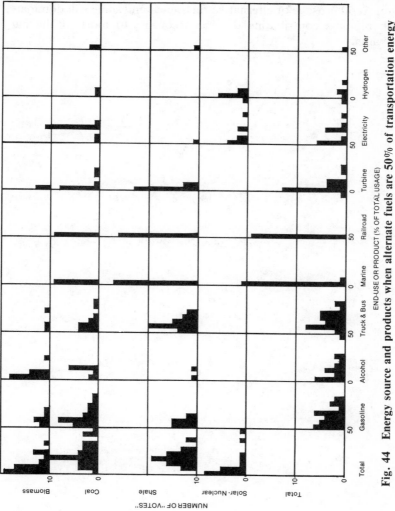

Fig. 44   Energy source and products when alternate fuels are 50% of transportation energy

biomass, and none from nuclear or solar. The total would be about equally distributed among gasoline, alcohol, and truck/bus.

(b) When alternative fuels reach 15% of the total transportation supply, solar and nuclear begin to appear, and their use grows proportionately when total alternative fuel reaches 50%.

(c) At 50% usage of alternative fuels, coal and shale still dominate, with biomass contributing 0-10%, according to most of the participants.

VII.
IMPACTS AND INSTITUTIONAL ISSUES

A. A General History of the Nebraska Grain Alcohol and Gasohol Program*

Charles R. Fricke
Agricultural Products Industrial Utilization Committee
Lincoln, Nebraska

The history of Nebraska involvement in blending agriculturally derived ethyl alcohol and gasoline has been extensive. The first widespread evidences of this idea date back to the 1930's when an alcohol-blended fuel called Agnol was sold and distributed in Nebraska and surrounding states. Another alcohol-blended fuel called Alky-Gas was distributed in Nebraska and surrounding states in the late 1930's and early 1940's. Generally, there was a great deal of interest at that time expressed by several states in this Midwest region in grain alcohol-blended fuels. Three consecutive national conferences were held in Dearborn, Michigan, sponsored by Henry Ford in 1933, 1936, and 1938 to discuss the widespread use and development of grain alcohol-blended fuels. Several interest groups from Nebraska participated in these meetings.

During World War II, three alcohol plants were established in the Midwest with the assistance of the Defense Plant Corporation. One of these plants was built in Omaha, Nebraska to assist in the manufacturing of fuel extenders and synthetic rubber for the war effort in the early 1940's. This plant was later scrapped in the early 1950's. During the late 1950's and 1960's, there were several efforts to promote the use of grain alcohol in automotive fuels. The primary thrust behind the movements, as it had been in the past, was to utilize surplus Nebraska agricultural crops.

As grain surpluses mounted and export markets were troublesome in 1968, the idea gained considerable strength in Nebraska. At this time, the Nebraska wheat growers became quite active in promoting

---

*Much of the technical and economic information that appears in this paper has been furnished by Dr. William A. Scheller, Chairman, Department of Chemical Engineering at the University of Nebraska at Lincoln, in various papers presented throughout the nation. This information and certain other research information that appears in this paper have been developed by the University of Nebraska-Lincoln through research funds provided by the A.P.I.U.C.

the concept of using wheat and other feed grains for conversion into ethyl alcohol. In 1971, several farm senators in the Nebraska legislature obtained unanimous passage of LB 776. This legislation was signed into law that spring. This law provided for the establishment of the Agricultural Products Industrial Utilization Committee to administer the Nebraska Grain Alcohol Program. Membership of the A.P.I.U.C. consists of four people actively engaged in farming, two in business, and one representative of the petroleum industry. The Committee's primary responsibilities are to: (a) establish procedures and processes to manufacture and market agricultural ethyl alcohol-blended fuels; (b) establish procedures to make the blended fuel marketable by private enterprise; (c) analyze the marketing process and testing of marketing procedures to assure acceptance in the private marketplace of such blended fuels and by-products resulting from its manufacture; (d) cooperate with private industry to establish privately owned agricultural ethyl alcohol manufacturing plants in Nebraska to supply demand for such product; (e) sponsor research and development of uses for by-products resulting from the manufacture of agricultural ethyl alcohol in order to enhance economic feasibility for alcohol; and (f) confirm the acceptability of the utilization of an alcohol-gasoline blend as an automotive fuel.

The ultimate goal of the Committee is to obtain the construction and completion of one or more ethyl alcohol plants in the state of Nebraska.

Funds to carry out these broadly outlined activities just mentioned are obtained by withholding one eighth cent from the gasoline tax refund which is otherwise retained to users of gasoline for off-highway purposes. The revenue generated by this tax measure has amounted to an average of $82,000 annually. Obviously, the majority of these funds is derived from Nebraska's farmers. LB 776 also provided for a reduction in the state gasoline tax of three cents per gallon on the first ten million gallons of Gasohol sold each year.

The Committee's first projects and studies began in 1972. Because of the small budget and structure of the Committee, its administrative staff is small. Therefore, the Committee awards grants for the necessary research and studies elsewhere to public institutions or private entities. One of the first things that the Committee did was to begin collecting all the available information on the subject of blending alcohol and gasoline. The Committee gained the extensive cooperation in this regard of the Department of Chemical Engineering at the University of Nebraska at Lincoln. Much of the cooperation was generated by Dr. William Scheller, who was and still is Chairman of this department. Dr. Scheller is credited with the origination of the trade-name "Gasohol" in the early days of the Committee's existence. The Committee then initiated several small projects investigating the technical aspects of Gasohol. One such project was conducted in 1972. This project was a small-scale road test with half-ton pickup trucks using Gasohol and regular gasoline.

In 1973, the Committee reviewed the results of this project. They were encouraged enough to have a large-scale automotive fleet demonstration test designed by the University of Nebraska at Lincoln. This road test was designed with the cooperation of the Nebraska Department of Roads and the Department of Chemical Engineering at U.N.L. Dr. Scheller was designated as the principal investigator in this road test, which is properly titled "The Nebraska Gasohol Two Million Mile Road Test." Dr. Scheller recommended that 200-proof anhydrous ethanol be used for the alcohol in Gasohol. Using this type of alcohol would avoid any water problems and provide good starting in cold weather. So, the official definition of Gasohol became 10% agriculturally derived (200-proof) anhydrous ethanol and 90% unleaded gasoline. The alcohol for this project and remaining projects was obtained from a Georgia-Pacific Corporation plant in Bellingham, Washington. Georgia-Pacific fermented wood wastes into anhydrous ethanol. This alcohol is chemically identical to alcohol fermented from grain. This alcohol satisfied the Committee's project needs, especially since there was no other major fermenter of anhydrous ethanol closely available at that time. The alcohol was shipped to a Farmland Industries Cooperative Refinery at Phillips-burg, Kansas, where it was stored and blended into Gasohol. The Gasohol was shipped to three Department of Roads test stations spread strategically across the state of Nebraska. The Department of Roads furnished 45 vehicles to be fueled with Gasohol, unleaded gasoline, and regular gasoline. The road test began in December 1974 and was completed in October 1977. This project was the primary and most comprehensive that the Committee had ever financed and undertaken.

The preliminary results of the Gasohol Two Million Mile Road Test are encouraging. Consumption of Gasohol was about 5% less than for unleaded gasoline. No unusual engine wear or carbon build-up was found. Drivers reported that they experienced no problems of starting, vapor lock, or drivability. The Committee obtained the cooperation of the U.S. Department of Energy on exhaust emission tests at the Bartlesville Research Laboratory. These tests show that Gasohol emitted one-third less carbon monoxide than unleaded gasoline on the Nebraska road test vehicles. Because of the scale and results of this road test, it gained considerable national recognition.

While the road test was in progress, the Committee increased its pace of activities. One of the most far-reaching Committee projects took place in 1975. An experiment entitled "The Holdrege Gasohol Consumer Acceptance Test and Marketing Survey" was organized and implemented during that year. After the first positive reports of the road tests were obtained, the Committee decided that the public acceptance of Gasohol should be tested. Arrangements were made with the Holdrege Cooperative Service Station at Holdrege, Nebraska to sell Gasohol. The cooperation of Farmland Industries, Inc., of Kansas City, Missouri was easily obtained for this experiment since they were already storing the alcohol and blending it with their unleaded gasoline for the road test.

This experiment proved to be extremely successful. What was to be a project spread over a period of nine months to one year turned out to last only 2½ months. The project began June 1, 1975 and ended August 19, 1975. The Committee sold approximately 93,000 gallons of Gasohol. The supply of Gasohol was insufficient to meet the demand of consumers. Customers reported increased mileage and performance with Gasohol. The Committee published the sales and consumer acceptance data in December 1975 after all of the information had been compiled and computed. The Committee concluded from this project that a market for Gasohol definitely existed in Nebraska.

The Committee dispelled the misnomer in 1976 that food would be taken away from starving people if Gasohol becomes widely distributed. This was done through the auspices of a National Science Foundation grant to the University of Nebraska at Lincoln for research on the extraction of protein from the distillers' dried grains of ethyl alcohol fermentation. This extracted protein, commonly called a protein isolate, can be used in human food. This development reveals the prospect that the world's supply of protein can be increased. This research indicates that when one makes ethyl alcohol from grain, recovers half of the protein from the distillers' by-products, and feeds the remaining residual grain to cattle, 50% more protein is available for human consumption than if the original whole grain had been fed directly to cattle. This research is the reason that the Committee adopted as its motto "Food and Fuel for the Future" for its grain alcohol program.

During the last two years the Committee has undertaken many other significant projects related to Gasohol and agricultural product fermentation. In the latter part of 1976, the Committee obtained the services of an outside engineering consulting firm to conduct a preliminary economic feasibility study based on a coal-fired ethyl alcohol fermentation plant producing 20 million gallons per year. The study, which was released in January 1977, revealed that there was a profitable return on the capital investment of approximately $21 million. The Committee conducted other research on the by-products of ethanol fermentation, such as feeding trials, with promising results. The Committee stepped up its efforts at this time to attract private industry into Nebraska to construct alcohol plants.

Dr. William Scheller, who also serves as the Committee's technical advisor, and four other gentlemen from Lincoln formed the Nebraska Grain Alcohol and Chemical Company after reviewing the information from the Stone and Webster study. Their main objective was and still is to construct an agricultural alcohol plant that will produce 20 million gallons per year. This move confirmed what Dr. Scheller had calculated for a long time: that such an operation was technically and economically feasible.

In the spring of 1977, the Nebraska legislature voted to raise the three-cent tax credit on Gasohol to a five-cent tax credit on the first 20 million gallons of Gasohol sold per year.

In May 1977, Nebraska members of Congress introduced significant amendments to the 1977 farm bill. This legislation provides for four loan guarantees up to $15 million dollars each for the construction of industrial hydrocarbon or alcohol plants in the United States. The amendment also provides for $24 million worth of grants to be allocated to land grant colleges and universities for research on the processes and by-products of hydrocarbon or agricultural alcohol production. The seed for this legislation was planted by the Nebraska Gasohol Committee. The U.S.D.A. is presently implementing this program rather slowly, much to the dismay of the midwestern agricultural sector.

The federal farm bill legislation contributed more than anything else in bringing together many midwestern states on the subject of Gasohol. Many states began to contact the Nebraska Gasohol office due to the legislation. Seeing a strong demand for this information, the Committee felt it was time to demonstrate its work on Gasohol for the benefit of the agricultural, automotive, and petroleum interests. The Committee decided to hold the Nebraska-Midwest Regional Gasohol Conference on November 1 and 2, 1977. The conference was expanded to incorporate reports on Gasohol-related projects in other states.

In August 1977, the Committee voted, in the midst of much controversy, to undertake a comprehensive economic Gasohol feasibility study recommended by Governor J. James Exon. Governor Exon obtained $30,000 from the U.S. Department of Energy and $30,000 from the Old West Regional Commission to finance the feasibility study. An outside consulting firm was selected to conduct the study. The study is scheduled for completion on July 1, 1978.

In the fall of 1977, a great deal of interest was generated in the U.S. Congress on Gasohol. Several Gasohol-related bills were introduced to provide for the use of Gasohol. A good example of this legislation is an amendment to President Carter's energy bill that provides for the deletion of the four-cent federal excise tax on all gasoline containing 10% alcohol derived from agricultural products and coal. This legislation was initially introduced and cosponsored by several U.S. Senators from the Midwest (principally U.S. Senators from Nebraska, Kansas, and Illinois). Congressman Glickman of Kansas introduced legislation calling for a nationwide mandate for 10% alcohol in all gasoline in the United States. Senator Church of Idaho introduced similar legislation in early 1978. All of this legislation is still pending as of this writing.

The Committee obtained the cooperation of the U.S. Department of Energy on a joint project in November 1977. The major objective of this cooperative effort was to "obtain estimates of (1) the cost to provide incremental increases in gasoline octane quality via the adjustment of refinery processing which should include the cost to refine a low octane no-lead base stock as low as 87 octane or lower; and (2) estimates of the value of the octane-boost quality of anhydrous ethanol when used as a motor fuel component. 'Value' as used in this Agreement would mean the change in fuel octane quality realized from addition of a given quantity of anhydrous ethanol to gasoline and the gain in automobile engine efficiency that is made possible with the readjustment of engine design and operating variables to utilize the octane quality increment as described above. Another objective of this joint effort will be to determine which gasoline additives anhydrous ethanol can replace, e.g., MMT and others." The Committee provided $10,000 toward this project. The project is nearing completion at the present time.

The Nebraska-Midwest Regional Gasohol Conference was attended by people from all walks of life. Representatives from 26 states and 3 provinces of Canada participated in the conference activities. The proceedings of the conference were recorded and published in a booklet in February 1978. The conference proved to be quite satisfactory and successful. The Committee is considering making this conference an annual event.

A significant resolution passed by the participants of the Gasohol conference recommended that:

> The Nebraska-Midwest Gasohol Conference go on record in support of the creation of a NATIONAL GASOHOL COMMISSION to coordinate research, develop resources, encourage public interest, and to provide public information on the production and use of grain alcohol, and invite grain producing states to join by participating through legislative, administrative, farm organizations and civic groups—AND that the NATIONAL

GASOHOL COMMISSION hold an organizational meeting by March 1, 1978 for the purposes of creating by-laws and establishing criteria for membership.

The resolution passed at the November Midwest-Regional Gasohol Conference prompted the Committee to take upon itself the initial task of organizing the first meeting of states interested in creating such a commission.

In December 1977, the Committee chose from a list of participants at the November conference temporary appointees to represent their respective states at the first commission meeting.

The first organization meeting of the National Gasohol Commission took place in Lincoln, Nebraska on January 24, 1978, at the invitation of the Nebraska Gasohol Committee. It was at this meeting that the Commission was officially formed. Representatives from 10 states participated in this formation. Representatives were asked to go back to their respective states and obtain funds, state appropriations or otherwise, to finance the Commission. A second meeting was held again in Lincoln on March 7. Representatives from five new additional states participated at this meeting. A total of 16 states aspire to become official members of the Commission in July 1978, when the Commission can officially conduct business. The basic purposes of the National Gasohol Commission are to: (a) coordinate and disseminate existing information on Gasohol and agriculturally derived alcohol fuels among the member states; (b) develop and promote state and federal legislative incentives for Gasohol; and (c) develop and coordinate any further research on Gasohol and agriculturally derived alcohol fuels needed so as to avoid research duplication among the member states. It is highly probable that the headquarters for the Commission will be located in Lincoln, Nebraska.

Probably one of the most recent significant developments has been the opening of a service station in Lincoln, Nebraska selling Gasohol. A partnership called *GAS PLUS* has been pumping about 8000 gallons of Gasohol per week since the end of February. They are selling Gasohol for 69.9 cents per gallon. This has been a very competitive price due to customer reports of increased fuel economy and performance. *GAS PLUS* and the company formed by Dr. Scheller are making plans to market Gasohol extensively across the state of Nebraska this summer.

Since the advantages of the Gasohol fuel have been mentioned, it would now be appropriate to summarize in broader terms some of the ways that grain alcohol and grain alcohol plants can be very beneficial. The list is as follows:

(a) Grain alcohol plants can provide an alternative market for surplus grains that flood the normal marketing channels, causing lower prices.

(b) Grain alcohol plants can provide an alternative marketing channel during the annual railroad boxcar shortage.

(c) Grain alcohol plants can provide a market-saving channel for spoiled grain due to moisture, insect damage, and other natural disasters. These problems are increasing due to the longer storage requirements in the farm program.

(d) Grain alcohol is a renewable source of readily usable liquid energy which can be used immediately by itself or in gasoline blends to extend decreasing supplies of crude oil.

(e) Grain alcohol could be readily used in an oil refinery system to increase the octane level of the gasoline refined. This system would save refineries money and extend their crude supplies.

(f) Grain alcohol is a nonpolluting additive for improving the quality of unleaded gasoline. Many newer automobiles do not perform well on unleaded gasoline.

(g) The importation of foreign oil could be reduced with the application of 10% grain alcohol in gasoline on a regional basis. The United States balance-of-payments situation could be improved with the use of grain alcohol.

The Committee is looking ahead in the next year or so to developing small feasible agricultural alcohol-producing units that can be located in small rural communities or on the farm. The Committee will be concentrating also on the utilization of farm stover and other forms of bionass for alcohol fuels.

These Committee projects and progressive thinking are the reasons why Nebraska is called the national leader of the Gasohol movement.

This concludes a general summary of the history of the Nebraska Gasohol and Grain Alcohol Program.

B.　The Seattle Experience:
Conversion of Solid Waste to Fuels or Chemicals

James K. Coyne
Coyne Chemical Company
Philadelphia, Pennsylvania

1.　Abstract

This paper discusses the efforts of the city of Seattle, Washington, the Coyne Chemical Company, the former Energy Research and Development Administration, and other public and private sector participants to develop a full-scale solid waste to methanol/ammonia plant. Nearly four years of analysis, research, and development ended in May 1977, with the decision to cancel the program. A series of significant economic, social, bureaucratic, and organizational constraints to the effective implementation of large-scale alternative energy programs is evaluated. The hypothesis is offered that the existing unwillingness of venture capital sources to accept the risks inherent in alternative energy development, coupled with the procedural difficulties of marshaling public sector acceptance and support, will conspire effectively to prohibit the completion of any major alternative fuel development programs over the near term.

2. Introduction

The so-called "energy crisis" and the new economic realities of the seventies have produced for those individuals interested in the general field of energy a new specialty: one I might half-seriously call "alternativology—the study of energy alternatives." The classics might have invented the term "thermoptology" to say the same thing.

The United States, indeed the world, is blessed (some would say cursed) to have tens, perhaps hundreds, of thousands of concerned researchers diligently seeking the best alternative energy answer from among thousands of likely candidates. With the transportation element of the global energy equation having such significance, there has been no shortage of creative and innovative proposals as to how we find substitutes for the limited supply of liquid fossil fuels.

And yet, there have been to date only a handful of cases where public and private resources have been successfully marshaled to develop, finance, and construct a full-scale, operational facility to produce market-quality alternative transportation fuels. The program that was attempted in Seattle over the past four years provides an excellent case study of some of the potential problems and promises to confront the next ambitious thermoptalogist trying to show forth his own special alchemy.

I must point out at the beginning, however, that the Seattle project took an early turn away from methanol, our proposed alternative fuel, to ammonia instead. The reasons for this critical choice will become self-evident, and, in hindsight, it may have been this crucial decision that led to the collapse of the project. Nevertheless, it is problematical at this point to conclude that the methanol route would have been any easier. The lessons are just as relevant in either case.

3. Stage One: Early Synergy

Most major research and development activities of the type we are discussing here can overcome their first financial and organization threshold only if there exists a set of participants with the requisite skills, each of whom is seeking to solve a meaningful problem for himself or striving for a significant individual goal with the concomitant rewards. In other words, developing something as complex as a production facility for an alternative fuel or chemical feedstock cannot, at least in our modern society, be done without a diverse set of individuals and organizations, each of whom is coming at the problem from a different angle for his own "selfish" reason.

Such, certainly, was the case with the Seattle projet. The city of Seattle was, and is, producing approximately 500,000 tons of municipal solid waste that the city was obligated to dispose of in the most economical and environmentally appropriate way. Continued landfilling in the environs of Seattle was felt to be less than environmentally optimal, and, ironically, the high cost of fuel had inflated the cost of transporting the waste to distant disposal sites. The city's political leadership was also seeking a creative solution that could allow it to handle this major municipal problem without becoming involved with the surrounding King County government. The alternative of "burning" the trash to produce steam or electricity, as had been successfully done in Europe and a few American cities, wasn't especially suitable to Seattle, where cheap hydropower was plentiful.

The private sector was coming at the problem from a different angle, but its goals were just as important. The energy crisis of 1973-74 had left the private sector, in general, reeling from the chaotic effects of a tripling of the cost of most fossil fuels. The transportation sector may have garnered all the headlines as half a nation queued up to buy gasoline, but the chemical industry was probably impacted to an even greater extent. And, of course, it was the chemical industry, in particular, that saw the promise in the Seattle project. Natural gas contracts were being rewritten with new escalation clauses that promised a tenfold increase in the cost of this critical feedstock. Ammonia spot prices climbed to $500/ton (versus $25/ton five years before). Methanol quotes of $1.50/gallon weren't uncommon—and only a few years before, it couldn't command a modest 30¢/gallon.

The federal government also had some objectives that had implications for the Seattle project. The newly created Energy Research and Development Administration (ERDA), the principal predecessor of our modern-day DOE, and the Environmental Protection Agency (EPA) both had small units charged with the development of energy from waste technologies. And several prominent legislators from the state of Washington were anxious to see a full-scale demonstration program in the Evergreen State. Nevertheless, federal support was forthcoming only at the very end.

This project, like most, began with a curious individual who persuaded those working with him to let him undertake a small feasibility study. The man was Robert Sheehan, an engineer working for Seattle's municipally owned electric utility, and by the end of 1974 he had produced a preliminary feasibility study that proposed to convert the city's half million tons of trash into $15 million worth of methanol or ammonia each year. The cost at this point: only $60 million. The city would pay almost $5 million to the project to take the trash off their hands, and the combined total income would easily cover amortization, operational costs, and even a bit of profit for the private sector participants. Such is the stuff that dreams are made of.

At the same time, dozens of chemical companies across the country were looking for alternative sources of chemical feedstocks, ammonia, methanol, and hundreds of petrochemical products whose production was sharply curtailed or facing allocation. Two in particular became involved with Seattle. The Union Carbide Corporation, through their Linde Division, had been developing a proprietary pyrolysis technique that could "gasify" hundreds of tons of municipal waste daily and produce a synthesis gas that would be suitable for the manufacture of either methanol or ammonia.

The Coyne Chemical Company, at the same time, was pursuing a strategy of developing small ammonia and methanol synthesis plants in locations close to nontraditional feedstocks, e.g., offgases from steel plants, agricultural wastes, biologically produced methane, etc. Solid waste, after undergoing the pyrolytic combustion of Union Carbide's system, would produce a low-Btu synthesis gas that, with the appropriate engineering, could be suitable as a feedstock for one of their small methanol or ammonia plants.

4. The Request for Proposals

In early 1975, Seattle undertook a national search for private sector participants in a solid-waste-to-ammonia/methanol plant. Ten proposals were received, several of which were simply engineering organizations submitting their qualifications, and out of these ten, the team of Union Carbide and Coyne Chemical was selected.

But, in fact, all that had been done at this point was to determine the members of the team. Only the barest understanding existed as to how the plant would finally be financed, what the city could actually commit to provide, who the end user of the chemicals would be, where the plant could be built, how operating costs might be shared, and what levels of profit, it any, would be allowed. There were hundreds of loose ends to be tied, and the engineering was really only slightly past the pilot plant stage. And yet, at this early stage, it was politically required to make the choice between methanol and ammonia. It was, after all, unprofessional, at best, to appear uncertain as to what we were really going to do with a half million tons of old newspapers and garbage bags. Besides, project leaders wanted to make as many decisions as fast as they could to keep the momentum going. So, the decision was made to make ammonia.

5.   The Ammonia Choice

The ammonia choice was selected for three very different reasons:

(a) The methanol market was comparatively thin and a major production facility of this size in the Northwest could succeed only if a market developed for methanol as a component in transportation fuels. This was construed as one more market risk on top of the already compounded technological, developmental, and operational risks inherent in the project. Additionally, going into the transportation marketplace was viewed as competing with the oil companies—an unappealing idea at best. At the time they were all universally opposed to any form of doctoring with the gasoline marketplace which, after all, was already in chaos.

(b) Ammonia prices were more favorable. Methanol pricing, it was felt, could only respond to actions in the gasoline pricing equation. Taxes, too, were a big questionmark. Ammonia prices had been comfortably above $200/ton for over eighteen months, and at these levels there was a reassuring $60/ton profit margin over expected break-even prices. Also, this plant could possibly be the price-determining factor in the region, although other large ammonia plants were under construction in Canada and Alaska.

(c) Ammonia production would be welcome by the farmers of Washington state, and by the legislators who represented them, as well. Eastern Washington was becoming one of the major agricultural growth centers of the country, and yet it was paying at least a 20% premium for ammonia (the principal synthetic fertilizer of the region) compared with the central part of the nation. Ammonia could be easily converted into liquid or solid blended fertilizers, shipped at low cost to farming centers, save the farmers money, and help the citizens of Seattle hold down the cost of trash disposal.

The factors that were overlooked are far more complicated. Ammonia had long been a world-scale commodity that followed disastrous macroeconomic supply-and-demand swings. The situation was exacerbated by the uncoordinated headlong rush of third world countries to build ammonia plants, where formerly natural gas deposits were being wasted or ignored. Farm prices also took a tumble, and farmers became more careful as to how much ammonia they used and what they paid for it.

None of these problems or issues were unforeseen or ignored. Yet the decision to produce ammonia was made, methanol was put on the back burner, and the track was laid for the next two years of project development. Ironically, it was learned about a year later that methanol was even easier to produce from the Union Carbide synthesis gas than ammonia. But at this point, too much had been committed to an ammonia choice to change horses. As methanol prices kept rising, and ammonia prices started to fall, there developed a sudden interest in microeconomic theory, supply-and-demand equations, and even hope that a few prayer breakfasts might change the persistent tendency of ammonia to cheapen itself in the Northwest U.S.

6. Market Development

Once the decision was made to make ammonia, it was clear that the ammonia market must be analyzed, buyers approached, contracts developed, and distribution arranged. If this had been a traditional private sector investment in a new ammonia production facility, that would have been enough. But in this case, with a complicated private/municipal financing tied partially to public revenues and public indebtedness obligations, the rules were vastly different. You didn't just have to know where the customers were and what they would pay; you had to get them to sign firm take-or-pay contracts for as long as you could get—just to be sure that you were always above break-even. To make matters a bit tougher, you had to get the customer to agree to take the risks that costs might climb and he might have to pay more. Such customers are not easy to find.

Fortunately, these were short times in the ammonia market and customers were anxious to find reliable supplies. Farmers in the Northwest paid an average of more than $300/ton for ammonia in 1975. We felt it could be made for $155/ton. We weren't using natural gas as our feedstock (which normally represents nearly two-thirds of the cost of production) so we felt we could offer a greater hedge against continuing rising ammonia costs than a producer that was relying on natural gas as his feedstock. The U.S. consumed almost 18 million tons of ammonia in 1976. The Northwest took almost 15% of that. Our 140,000 tons of annual production, we hoped, could quickly be sold under terms that satisfied the politicians, the bankers, and the lawyers.

We quickly found four major users in the region who together bought 500,000 tons/year. And, they felt our price was fair—as a spot price. But they, more than we, could remember the wild fluctuations in prices when wheat prices fall, or when the Canadians start exporting, or when major new international plants come on stream, or when tankers become cheap to charter.

We responded, of course, that cheap energy was a thing of the past. Times were different now. It was the alternative energy source that was the feedstock of the future. Russians needed more and more wheat. The age of plentiful food and fuel was past, and renewable resources were the economic long-term way to go.

Then, of course, we appealed to other than economic sensitivities. This was good for the country, good for the Northwest, good for Seattle, and good for them. After all, they couldn't get as favorable prices long or short term from anyone else (of course, no one else would even talk long term, then). We then felt obligated to sweeten the deal, and added a back-up naphtha reformer to supplement our feedstock if the trash ever vanished. We were then the only two-feedstock ammonia plant in the country, even if we were only on the drawing boards.

Finally it dawned on us that they, the marketplace, knew something we didn't know. They knew that we needed them, and they didn't need us.

My purpose in presenting this marketing saga so vividly is not to tantalize or overwhelm anyone with the cold, cruel realities of the energy marketplace. Rather, I hope to present straightforwardly some of the characteristics of the major energy consumer, either fuels or feedstock, which will continue to slow the development of alternative energy projects.

The buyer of major amounts of fuels, feedstocks, or energy-intensive chemicals is fundamentally a single agent acting intelligently and conservatively to deal with three fundamental concerns: (a) pricing risk, (b) supply risk, and (c) time risk. Any interest he might personally have to advance on important national development project for the general social good will inevitably be subordinated to these primary concerns.

Pricing. Pricing is the most obvious language of the marketer, but also the element of analysis with the greatest degree of uncertainty. We were trying to sell 25,000-ton annual contracts of ammonia in a marketplace where prices twice as high were being paid on the spot market. A purchase of tankcar quantities of transportation fuel would probably be quite similar. The problem in either case is the buyer wants to be sure what price he will pay, and, of course, he doesn't want to be at a competitive disadvantage to his competition. The seller of fuel or feedstock from the alternative energy plant, be it biomass, solid waste, coal derivatives, or whatever, can't really be sure what the price will be. He has three levels of uncertainty: How much will the full-scale plant really cost? How much will it really make? What will the operating expenses be? A developmental project that has any one of these uncertainty factors controlled within 10% limits is, in my experience, superb. The combined effects of these uncertainty levels can be pricing inaccuracies of as much as 50%. But even if the seller can calculate the selling price accurately, there is no way he can predict the competitive response he will face in the marketplace. Calculator prices, transatlantic airfares, and even "gas price wars" are examples of unexpected competitive pricing strategies that are designed to drive unwanted competition out of the marketplace. We, in Seattle, had no way of knowing what the traditional ammonia producers in the Northwest would do when our plant came on stream. The profits they were making during 1974, '75, and '76 could easily subsidize a few years of losses to guarantee that we never break even. But even if they never did something like that, just the realization that they might caused our customers to panic.

Supply. Production problems, quality problems, labor difficulties, distribution bottlenecks, feedstock, and raw material shortages: these are the parameters of supply risks. The major consumer of fuels or feedstock considers these the essential tradeoffs to price. Cheap ammonia is available from the Russians, but will it always be there? There is a small, old plant in New Mexico, but can it keep producing? You have a state-of-the-art plant on the drawing boards, but how long will it take you to debug it? You are using potatoes as your raw material, but what if there's a blight, a shortage, or no one to harvest them? The supply risk is the craps game of the fuel and feedstock production plant. Once again, the prudent buyer wants all his bets covered. Usually, this is accomplished by contracting with several diverse suppliers. "You get 20% of our business, and if things work out well, we might increase it to 40%," say the careful contract negotiators. We on the marketing side, of course, wanted to develop as large a contract as we could. Such was not our good fortune in the Seattle project. Furthermore, our technology was so new, i.e., producing ammonia from something other than natural gas (even though the early ammonia plants all used a synthesis gas similar to ours), that we had to demonstrate that we would add failsafe on top of failsafe to ensure continuous output. We needed redundant reactors, redundant gasifiers, redundant oxygen plants, redundant storage, redundant transformers, and finally even a redundant feed-stock—naphtha in permanent storage in case Seattle ran out of garbage or it couldn't be collected. The end result was that we had to approach and develop three times as many customers as might otherwise have been the case, and the capital cost of the plant had another $50 million of redundancies added.

Time. Whoever wrote the words "Time is on your side" wasn't talking about market development for an alternative energy plant. Time, it seems, is on the side of today. Tomorrow sells at a tremendous discount. The problem, of course, is that the price and supply uncertainties involve so many probablistic elements that the cautious buyer is unwilling to leave himself out on tomorrow's limb if one of the uncertainties turns against him. Tomorrow always means the worst. Your costs will be higher tomorrow, your competitor's price will be lower. The customer might not need your product tomorrow, or there might be a new technology that does your process even better. And all these problems are worse for the developmental promoter. Fundamental to the cautious buyer's makeup is an unwillingness to go it alone. If he rejects the wonderful new technology you offer to make jet fuel from turnips, and everyone else does, too, even though you tell him jet fuel is going to quadruple in price and go on allocation— he won't care, because he'll be in the same boat with everyone else, sinking or swimming with whatever the traditional jet fuel market does. Facing this inevitable resistance to go it alone, you can expect the marketplace to reject almost any contract that increases its risk over time. Escalation clauses are the classic case. Traditional ammonia contracts today include escalation clauses covering the price of natural gas. This can mean that as much as 65% of the selling price is vulnerable to increases dictated by the Texas Railroad Commission or some other regulatory body controlling gas prices (or, perhaps, someday, the marketplace forces of supply and demand). But if you offer an alternative plant making ammonia from trash, potatoes, or seaweed and ask to have escalation on only 20% of your costs, electricity and labor, for example, you'll more than likely meet with deaf ears. The cautious buyer doesn't want to find himself being escalated on one thing while the rest of the world is marching to a different escalatory tune. The end result of all of this was that the only "long-term" contract we could negotiate was at a price of $80/ton (compared with more than $300 in the field)—and this was only for ten years. The other buyers insisted on a no-lose contract. It would contract for a percentage of the annual requirements at 85% of the price being paid to everyone else. Sadly, this is likely to be the only major contract terms the alternative energy salesmen will be able to find.

7. Engineering

I don't think I have to point out that the engineering-and-design stage of a major alternative energy development project is even more complicated than the product marketing. Our process in Seattle encompassed four steps: waste collection and preparation, gasification, gas shift, and ammonia synthesis. There were dozens of complicated subprocesses within each major element, some of which were well-established, some only in the demonstration stage, and a few still at the pilot plant level.

Engineering is a struggle because everyone expects it to be so precise, when, in fact, it suffers from as many layers of uncertainty as the marketing and finance. But more specifically, there are three essential difficulties: coordinating all the teams, meeting public requirements, and determining the yield.

In the case of our effort in Seattle, at one point or another there were four sets of engineering consultants, the city's own staff, and analysts with the federal government. Getting agreement among them was difficult, at best.

The public requirements encompassed the normal environmental impact requirements, zoning, city and state permits, dozens of hearings, and hundreds of approval meetings. Everyone from the F.A.A. to the Puget Sound Commission had a time-consuming approval process to tackle. The public hearings, of course, had their advocates. Ammonia can be *accused* of being worse than a nuclear bomb testing facility. It would cause cancer, destroy the ozone layer, blow up, lower real estate values, dilute the bay, raise electricity rates, increase traffic, or interfere with FM radios. It wasn't relevant that it would dispose of garbage, save the countryside from landfills and long-haul garbage trucks, save natural gas for better uses, reduce agricultural costs, improve the balance of payments, or save the citizens of Seattle money. Public hearings are not intended to be balanced.

Fortunately, we passed all these tests, but the time costs were staggering. And the hardest part, it seems, is translating from the language of engineering to the language of public meetings without sounding condescending or defensive.

Everyone looks to the engineer for the proof that the bottom line makes sense. The public asks the chemical engineer to determine on the spot how much it will save in monthly taxes in 1985. The investment banker wants to know if expected down time will be 2.25% or 2.31%. The accountant has to know the construction cost inflation forecast to an accuracy of $\pm 0.25\%$. But most of all, everyone wants to know what the yield will be—how much will it make.

In the case of our process, this answer was the worst exercise of all. To begin with, we had a variable and seasonal amount of garbage, with a varying hydrocarbon and water content, which, in turn, would be gasified into a hydrogen- and carbon-monoxide-rich gaseous mixture with varying conversion efficiencies, which then was shifted to a different carbon monoxide and hydrogen blend, again with an unknown conversion factor, which, in turn, was to be synthesized into ammonia—hopefully following stochiometric ratios which were reasonably accurate.

The engineer, naturally, works with carefully proscribed assumptions which he dutifully attaches to his estimates as subscripts and footnotes. Assume, for example, that we continue to have 81.7% of the waste input stream composed of paper and plastics; then we can expect the following hydrogen and CO yield. Of course, don't point out that the state is considering legislation that would ban non-returnable containers. Similarly, scaling up several major components from pilot plant to full scale can be a hazardous exercise. In short, the engineer did not have an easy job.

Through all the chaos and uncertainty, we carefully projected our yield, coordinated our teams, and held the public meetings. And at the end, our expected yield was very close (within 10%) to our original projections. Unfortunately, the market price was collapsing and the capital cost was climbing. It was this latter problem that the engineers seemed most unresponsive to.

The inflation we all saw between 1974 and 1977 is, of course, quite memorable. It places unbearable burdens on the estimating profession. We were no exception. No one really knew if the 15-20% annual increases in construction costs would continue over the three-year construction period we foresaw from 1978-1981. The process industry was among the worst examples, and the engineers and estimators are naturally hesitant to come out on the short end, especially since they had been so soundly criticized in recent months for missing the mark so badly on other projects. Add to this the natural engineering tendency to gold-plate and overdesign, especially on highly visible, publicly supported projects like this. The end result was an escalation of the capital cost from $60 million to over $200 million, of which one fourth represented escalation and contingency costs alone.

Thus we evolved into the worst of all possible worlds—rising costs and falling prices.

8. Financing

How was all this to be paid for? Well, at first it was going to be easy. Originally, Seattle expected to finance approximately two thirds of the project with municipal revenue bond financing and one third by a private financing. It was naively expected that investment banks would accept the terms of the ammonia marketplace and the marketplace would accept the terms of the bankers.

Rather quickly it developed that such was not the case. The investment bankers for both the private and public sectors insisted on full risk protection—and not just market risks, but all operating and raw material risks, as well. In other words, the banks didn't want to be left holding the bag if (a) the plant didn't work, (b) there wasn't enough garbage, (c) the market price wasn't enough, or (d) output was deficient.

In our society this has become a common dilemma of recent times, either because of increasing uncertainty in our world, new timidity on the part of bankers, or a feeling that Uncle Sam is the risk-taker of first and last resort. Some say it's because our tax structure discourages the entrepreneur. Whatever the reason, the whole project turned to the U.S. government to guarantee the risk.

The government, oddly enough, were easy people to work with—at first. Washington State had powerful legislators with brilliant, tape-cutting staff. ERDA and EPA were eager to get something accomplished (both had had some embarrassments in this general area). The concept of loan-guarantees was gaining momentum in Congress.

The problem was that everyone felt that the solution was at hand: the generous Uncle would accept all risks and guarantee everyone's success. It was like Dorothy, the Lion, Scarecrow, and Tinman all blindly believing that the Wizard was a wizard who would serve. Furthermore, everyone became a bit less willing to accept *any* risks since, after all, they didn't have to. Also, everyone wanted just a bit more profit and an extra little cushion against any contingency. Thus, the engineers added more gold, the bankers took another half point, the consultants added another study, the city cut the disposal fee a little, and the accountant covered himself against inflation a bit more securely.

The result was that expected operating costs increased as dramatically as construction costs had. The required ammonia price climbed to $180/ton. The government was now faced not with the risk of having to bail out the project, but the certainty of doing so.

The project died.

Of course, it is easy to blame the banks, because, after all, if they had loaned the money, taken a chance, it would have been a daring and, some say, brilliant move. Construction costs have not climbed as fast as was feared, and ammonia prices have stabilized around $150/ton. The city is paying stiff prices to haul away its trash, and the numbers probably would have worked—and the banks could have been heros.

Unfortunately, bankers are no longer in the business of being heros, especially on big visible projects. Collateral has to be there, and the mercurial ammonia market is not good collateral, even if the city of Seattle is a much better credit risk than New York. The days of effective venture capitalization are gone, at least for now. The public equity route has disappeared. REIT's burned too many lenders to let them be forgotten. Sure, there are tremendous amounts of cash floating around, but the cash drains of the federal government deficit, an export merchandising deficit, and booming consumer credit more than soak up the excess. Only recently have any significant amounts of new cash moved towards low-risk windows on Wall Street. The venture capitalists have no funds.

We are sadly aware of the great alacrity with which the government has moved to develop an energy bill which provides at least some relief for this problem. The loan guarantee and commercialization programs are still prefetal. Big projects, especially those with Big Oil or Big Chemical participants, are least likely to see the government taking a realistic view of risk. And the picture doesn't look good.

We are in the midst of a era of finger pointing. Government points to industry, industry at banks, and banks turn to government. To me it seems that our system must develop a new awareness of the need for enterprise and the entrepreneur. This is not just the individual—but the large corporation, as well. If things that need to be done, like developing alternative fuels for transportation or new feedstocks for fertilizer, are not getting done with the "system" of public and private cooperation (disoperation?) that exists today, then a change is needed: a change in attitude, a change in government incentives and disincentives, a change in education, a change in our financial institutions. We need more free enterprise and more free entrepreneurs.

9. *Conclusion*

The Seattle project to produce for the first time a major energy-intensive chemical from a renewable resource, solid waste, failed. It failed despite the fact that the public and our legislators were tremendously impressed by its promise. It failed despite the fact that hundreds of brilliant men and women were spending thousands of man-months and over $1,000,000 to prove that it was a more intelligent way of making ammonia (or methanol) and a better way of recycling the hydrocarbon component of our mountains of solid waste. It takes nature 60 million years to recycle a plastic garbage bag. We could do it in a few hours.

The Seattle project failed despite the fact that many men and women supported it, and despite the fact that it solved many problems, and despite the fact that it was less risky than the first natural-gas-based ammonia or methanol plant. It failed, despite the fact that in many other countries it would not have failed.

It failed because of inflation. It failed because no one will accept the risks of inflation. It failed because inflation kills enterprise and the entrepreneur.

And that may be the reason that many of your projects will struggle over the years ahead. Oddly enough, inflation in many respects gives us all our reason for being here today—the inflation brought upon us by the tripling of the price of oil. And yet, inflation will make it almost impossible to get your project off the ground unless our society comes to grip with inflation's risk. I am pessimistic, but you don't have to be so.

C. Synthetic Fuels Program

Charles L. Stone
California Legislature

1. Introduction

"What do we do when the wells run dry?" Mr. Dan Rather of CBS asked that question two years ago as part of a *Sixty Minutes* program regarding methanol. The answer to that question is synthetic fuels. Which fuels, ethanol, methanol, or synthetic gasoline, is really not the issue. We probably can and will, out of necessity, use all of these fuels, and others, too, as the oil and gas run out.

The California Legislature has been pressing forward for a number of years to bring about the introduction of all kinds of synthetic fuels. One essential fact we discovered, as we developed our approach, is the fact that waste materials, biomass farming, and coal are essential basic resources for producing any synthetic fuels we may end up using.

We have, therefore, set out to do certain production and utilization demonstration projects which will give us the confidence to support the building of a complete synthetic fuels industry. We consider that industry to include capturing the sun's energy through photosynthesis; converting resultant plant life into synthetic solid, liquid, and gaseous fuels; and providing us with machinery which is designed to be compatabile with those synthetic fuels. Energy and dollar efficiencies are expected to change rapidly as commercialization becomes a reality. Academic paper studies and opinions will give way to actual practice and real experience. Doubts and apprehensions will give way to confidence and satisfaction once commercial practices are well established.

2. Energy Freedoms

The citizens of the United States have come to expect energy freedoms as they expect other freedoms. They do not desire to give up their personalized mobility, their individual homes at 72°F, or their current chemicals and materials made from natural gas and oil.

Throughout history, the shifting from wood to coal to oil and gas, as a means of energizing machines, has introduced greater energy freedoms with each shift. Now we are about to experience a new shift—to synthetic fuels—which will undoubtedly become better than fuels of the past. The public will demand that they be better fuels, and modern technology will provide us with the means—once we have the will.

California could well produce all of its ground transportation fuels (35% of energy market) from biomass in twenty-five years. This goal alone would require the production of 24 billion gallons of alcohol fuels from about six million acres of biomass farms on marginal lands. Other materials could be used in lieu of that biomass farming material. The use of biomass farming, coupled with wastes and some coal, could provide a complete replacement for all the oil and gas normally removed from the ground. We are talking about a full synthetic fuels industry with large and small plants doing selected functions. Many jobs and new small-business opportunities are expected to result from this new industry. That is our expectation as we develop the California Legislature's Synthetic Fuels Program.

3. Alternative Fuels for Transportation

The subject of this Symposium is synthetic fuels for transportation. Of special interest to this group is the methanol-X car which the Legislature has produced as a prototype for future government fleet legislation.

The synthetic fuels car is being built and tested under an Assembly Rules Committee authorization. This methanol-X car is to be used as the prototype of a fleet of cars which the Legislature will be asked to support in 1979. One major concern we have is the availability of methanol-X to support the first fleets. The current requirement will probably have to be satisfied by alcohols from natural gas conversion. Residual methyl fuel may be available from refineries or chemical plants for as little as 5¢ per gallon. The fuel quantity and quality will be in question, however, if we use residual sources. Since we expect this car to provide a 135% energy efficiency compared to gasoline, it is our expectation that alcohol fuels from wastes, coal, and biomass farming will be readily available in the post-1985 period. Should we build the fleet cars first, or the methanol-X production plants first? This chicken and egg issue is frequently viewed as one reason alcohol will never be used in this country without government support in the beginning.

That is the reason the Legislature must be active in the promotion of a complete synthetic fuels industry now. Two bills we are currently considering are AB 3555 and SB 1662. These bills allow for an interim Gasohol blend approach to get a market started for synthetic fuels capable of being used in today's cars. Once these cars are on an optional gasoline-alcohol blend, we will be able to obtain enough synthetic fuel to support the first fleet operations on 100% synthetics. You might say we are going to let the gasoline industry help us hatch our first few eggs.

The gasoline producers who cooperate will be better prepared to convert to the synthetics as they move into the public marketplace. Those oil companies with coal holdings will undoubtedly be wanting the synthetic fuels industry to help them hatch a few of their coal mine eggs. Obviously, society will be better served by this cooperation between biomass and coal sponsors than if they work separately.

Most probably California will use a 50-50 blend of, respectively, charcoal from biomass and coal from other states, to provide synthetic natural gas and alcohol fuels to replace all the oil and gas well products now consumed. The characteristics of a cofeedstock, coconversion, and coproduction process give new, improved net energy and cost numbers. When coupled with gas enrichment during the gaseous phase, we can expect to produce pipeline quality gas and methyl fuels at less than $3.00 per million Btu.

The California Legislature passed a law in 1976 calling for a mobile pyrolytic converter to be built and in operation by July 1, 1979. The California Solid Waste Management Board is now working on getting that prototype unit built to the Legislative mandate. That unit will move from site to site, wherever large deposits of agricultural waste have been accumulated. By converting the waste into charcoal at the waste site, the logistics costs for hauling the material are reduced by approximately 75%. The charcoal is rated at 12,000 Btu per pound ($1.25 per million Btu) and is sulfur-free. When mixed with the 11,000 Btu per pound coal ($1.20 per million Btu), which is high on sulfur, we get a higher quality feedstock for gasification.

Within the next decade, California could have more than 400 mobile units traveling from farms to saw mills, to the forests, and then back to the farms. Each stop would provide for conversion of perhaps 30 or 60 days worth of waste at 100 dry tons per day, into clean-burning, fuel-grade charcoal at less than $1.25 per million Btu.

Each unit is self-propelled, self-powered, and independent. At $500,000 per unit, small companies or individuals may own any number of units. This new source of energy will prove to be cheaper and cleaner than any coal mined in America today.

Alcohols suitable for methanol-X type cars could well be produced on a commercial basis for 18 to 21 cents per gallon at the gate from the charcoal/coal conversion plants. Because of the increased (135%) mechanical efficiencies of the methanol-X car, we can expect to go further, quicker, with less emissions, at a lower cost per mile and less energy when the wells run dry.

Any movement toward development of a synthetic fuels industry will not mean an end to oil and gas well activities within the United States. Hopefully, however, it will mean that energy independence from foreign sources will be a reality as the world demand for natural oil and gas exceeds the world production in the late 1980's.

By moving now to establish the new alternative, we can avoid the social trauma and economic disruption associated with a shortage of solid, liquid, and gaseous fuels.

4.   The Methanol-X Car and Gasohol

It is necessary to focus the features of our proposed 100% synthetics methanol-X approach and the blend approach, commonly referred to as Gasohol, in order to understand why both will be needed. The technical difficulties of using methanol in a blend with gasoline are relatively easy to resolve. Methanol is also going to prove to be much cheaper than ethanol.

However, it is not clear that the cost of solving the problems of methanol blends (i.e., special bonding agents, materials compatibility, etc.) and the large-scale plant financing requirements, construction lead time, and current automobile designs will be able to compete with the ethanol approach in the next decade.

Ethanol manufactured in small-scale plants is looking more attractive for the short haul because of these technical and economic issues.

On the other hand, ethanol will not be able to compete with the higher production capability of large-scale methanol plants as envisioned by the legislative sponsors and shown on Figure 45. As the methanol productivity based upon biomass charcoal and natural coal emerges (post-1985), gasoline will begin to diminish as a promising main fuel for transportation.

Vehicles similar to the methanol-X car will be well established as fleet vehicles and general public uses will begin to appear. During the 1990's, ethanol will be used to blend with both the gasoline and the methanol-X fuels. Because of the high octane ratings of the methanol-X fuels, the cars designed to be using methanol-X will be operating at 14-to-1 and higher compression ratios. This means that the synthetic-fuel vehicles will not be able to use the Gasohol or gasoline fuels.

For the present, the methanol-X car is serving the Legislature as a tool to establish the art of the possible use of synthetic fuels in automobiles. The vehicle has been modified by a piston redesign, fuel injection, a new camshaft, and instrumentation.

A fourteen-month test program will be underway next week to baseline the performance, emission, and mileage potentials possible with minimum modifications. This effort will set minimum standards which can certainly be improved upon by the technical and financial resources of the government agencies and private industry.

New improved fuel formulas, based upon the captive talents of the oil industry scientists and engineers, will begin to emerge in the late 1980's. The automotive companies will be competing openly in the late 1980's to capture the emerging market for the vehicles capable of performing on the new fuels. Once this competition gets underway, the new energy freedoms of the synthetic fuels era will receive support

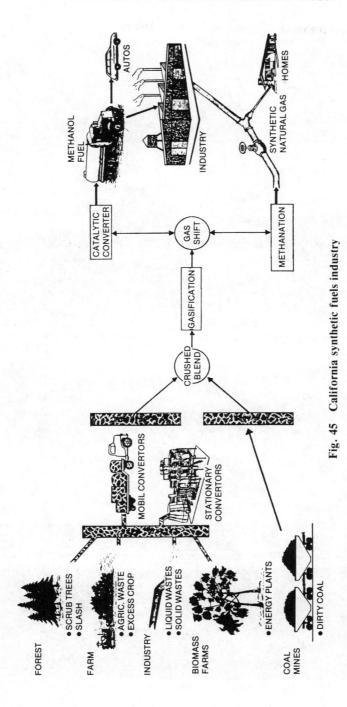

Fig. 45 California synthetic fuels industry

Fig. 46   California Legislature synthetic fuels program test car MX100

Fig. 47   California Legislature synthetic fuels program test car UG100

Fig. 48 Dashboard of the MX100

Fig. 49 Modified engine of the MX100

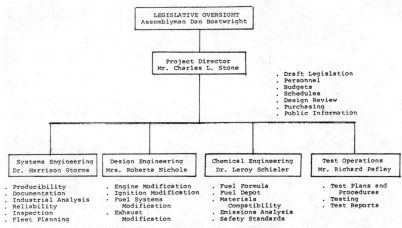

Fig. 50   Key personnel involved in project

from the auto and oil companies rather than the resistance they have both offered in past years.

Figures 46-49 show some of the features incorporated in the methanol-*X* car. Figures 46 and 47 show the two vehicles scheduled to be tested. The MX100 is the 100% methanol-*X* car and the UG100 is the 100% unleaded gasoline car. Figure 48 is the dashboard of the MX100 showing the tachometer, ampmeter, oil meter, temperature meter, and four pyrometers installed for engine monitoring. Figure 49 shows the engine of the MX100 as modified. A detailed technical report which will describe the first, second, and final configurations, as well as test data and procedures, will be published in late 1979. The methanol-*X* formula, which is being developed during this test period, will be fully explained in that report also.

The key personnel involved in this project are identified in Figure 50; projected fuel economy and emissions are shown in Table 34.

D.   Riverside's Hydrogen Bus Experience

Robert M. Zweig, M.D.
Pollution Control Research Institute
Riverside, California

1.   Abstract

Riverside's hydrogen bus is a symbolic example of community political pressure and government response. This joint effort was initiated by City Advisory Commissions and Chambers of Commerce, then encouraged by city, county, and state support. Past and present air pollution control measures in the South Coast Air Basin have been ineffectual, prompting this novel approach—development of a nonpollution vehicle—a hydrogen bus. Citizen concern regarding

Table 34 Synthetic Fuels Car: Fuel Economy and Emmisions[a]

| Automobile—Engine Description | Fuel Economy[b] | | | | Emissions | | | | | |
	Miles/Gal Urban[c] Hwy.		% Inc. over OEM Urban Hwy.		NOx g/mile % red.		UBF g/mile % red.		CO g/mile % red.	
Stock System—Gasoline	20.0	25.1	0	0	4.9	…	2.2	…	16.7	…
Stock System—Methanol	22.2	27.8	11	11	2.0	59	1.1	50	12.5	25
Improved Carb.—Methanol	23.2	29.5	16	13	1.8	63	1.5	32	12.0	28
Advanced Engine (10:1 CR Improved Carb. and Cam)	25.1	29.6	25	18	1.9	61	1.5	32	11.0	34
Predicted Results (14:1CR Engine—Methanol X)	26.5	33.7	32	34	1.0	80	0.4	81	7.0	58
1980 Standards	20 (comb. cycle)		…	…	1.0	…	0.41	…	7.0	…

[a] Based on simulation of the federal test procedure using engine-dynamometer test data at fixed equivalence ratio of $\Phi = 1.0$.
[b] Fuel economy in miles/gallon gasoline equivalent (energy basis).
[c] Urban cycle results adjusted for cold start.

public health effects of deteriorating air quality stimulated participation in experimenting with this new fuel. Riverside's municipal
fleet had previously been powered by propane fuel; thus a hydrogen
vehicle seemed to be the next logical step toward a cleaner gaseous fuel
operation. Improvement of air quality will result in cost savings for
pollution-related disease care. Further expansion of hydrogen
utilization for more fleets would further eliminate pollutants presently
produced from fossil fuel combusion. An added bonus of a
"universal hydrogen economy" would be independence from foreign
oil blackmail for our transportation industry. Local production of of
hydrogen as a replacement fuel would make a community, state, or
county "energy-independent." Hydrogen could be considered the
panacea for our problems of air pollution energy shortage and foreign
balance-of-trade economics.

2. Introduction

Riverside citizens requested a new approach to the smog problem
after experiencing twenty years of deteriorating air quality. In spite of
strategies mandated by federal (Environmental Protection Agency),
state (California Air Resources Board), and regional (Air Quality
Management Board) pollution control agencies, certain areas in the
South Coast AirBasin have been experiencing degradation of anbient
air quality. Several Health-related organizations had expressed a fear
that we may be approaching a catastrophic "killer smog" if a new
solution to the problem were not implemented.

The Environmental Protection Commission of Riverside requested
that the City Council address the smog problem by sponsoring a
"clean vehicle program." Preliminary trials evaluated methanol,
methane, propane, and diesel fuels, but the cleanest fuel was found to
be hydrogen. Consequently a request was made to the California
Assembly Transportation Committee for financial assistance to
develop a hydrogen Dial-A-Ride bus to add to Riverside's existing six-
bus diesel fleet.

An advisory committee consisting of members representing Jet
Propulsion Laboratory, UCLA, Union Carbide, Caltrans, and the
California Energy Commission deliberated and concluded that, for
safety considerations, an iron titanium (Fe-Ti) hydride storage system
should be utilized. Bids were obtained from four contractors, with
Billings Energy Corporation being awarded the contract. Delivery of
the bus took place in Riverside on September 19, 1977. Dr. William
Van Vorst, Chairman of School of Engineering at UCLA, was the
principal guest speaker at the dedication ceremony. The bus is an
Argosy Air Stream, 19-passenger minibus with a 455 cubic inch
Chevrolet engine. Engineering details are available from Billings
Energy Corporation.

From data obtained at Air Resources Board Testing Laboratory, we were able to calculate a 98% reduction of tailpipe emissions compared to a similar gasoline-powered engine.

Obviously, hydrogen-fueled vehicles should produce no toxic pollutants except for small amounts of NO_x. It is estimated that this would be less than 10% of EPA requirements for 1985. Water induction technique has resulted in a marked reduction of NO_x compared to earlier models of hydrogen vehicles.

Riverside is particularly interested in reducing air pollution as a means of improving public health. Several pollutants have been analyzed at the State Air Pollution Research Center, located at the UCR campus, and some of those levels are reaching dangerously toxic concentrations at various locations in the South Coast Air Basin. The following is a medical review of chemical components of smog as found in the South Coast Air Basin.

3. Health Effects of Pollutants

Carbon Monoxide. CO reacts with the hemoglobin molecule in our red blood cells, forming carboxy-hemoglobin, which reduces the efficiency of this oxygen-carrying molecule. As a result, small blood vessels react by constricting (vasospasm) from acute exposure, or by causing hardening of the arteries (arteriosclerosis) from chronic exposure. Those blood-vessel changes can result in a decreased blood supply to various organs. If the brain is affected, a stroke (cerebral vascular accident) will result; if the heart is affected, a heart attack (myocardial infarction) will result; if the kidneys are affected, high blood pressure (hypertension) will result. Increased levels of carboxyhemoglobin will also cause inefficient function of the central nervous system, resulting in a lack of discrimination of small differences of time and space, which can result in poor coordination.

Sulfur Compounds (SO_x, SO_3, SO_4, H_2SO_4, and aerosols). In toxic concentrations, SO_2 can cause irritation of the respiratory epithelium and spasm of the bronchioles (small airways). SO_2 and particulate combination caused the high death rate in London in 1952. It has been estimated that SO_2 produced from each coal-fired electric generating plant producing one gigawatt per year will cause 50 deaths per year. Acid sulfate aerosols are shown to be the most toxic oxide of sulfur (see Figure 51.)

Total Particulates. TP or smoke result from incomplete combustion of hydrocarbons; depending on the size and composition, particulates can cause many deleterious effects on the lining (respiratory epithelium) in the nose, sinus, throat, and lungs. When particulates coat this lining, there results a decrease in the efficiency of oxygen exchange; and, in combination with other compounds, these par-

Fig. 51   Sulfur dioxide levels in Riverside

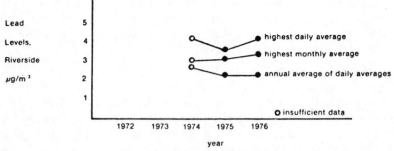

Fig. 52   Lead levels in Riverside

ticulates can cause tissue destruction, and, in higher concentrations, eventually can cause cancer.[102] Other pathways of cancer-producing pollution are presently being studied at various institutions throughout the country.[103] Nitrosamines,[104] formed from NH_4 and NO_x, are known to be potent carcinogens: ozone is shown to cause chromosomal breaks which are thought to be carcinogenic, cocarcinogenic, or mutagenic. Epidemiological studies suggest a definite relationship between fossil fuel production,[105] refining, and utilization and increased incidence of respiratory tract carcinoma.[106]

Lead. When lead is inhaled as the particulate, it is absorbed through the respiratory epithelium and terminates in the red blood cells and bone marrow. Certain concentrations in the body can cause serious problems in the bone growth of children and will also affect red blood cell formation and function. When in contact with nerve cells, it can cause loss of function of the nervous tissue.[107] Forty percent of the inhaled submicronic particles are retained in the body. Lead is difficult to remove from the body, and the excretion time can last several years. Ambient lead levels are rising in Riverside (Figure 52). Children living in Riverside have shown higher lead blood levels than the average California child.[108]

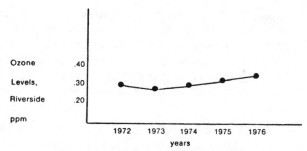

Fig. 53   Oxidant levels in Riverside

POLLUTION FROM PROPOSED HYDROGEN ENERGY SYSTEM

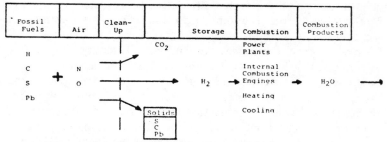

Fig. 54   Human exposure in the hydrogen economy

Photochemical Oxidants [*Ox, ozone* (*Ox₃*), *nitrogen dioxide* (*NO₂*), *and peroxyacetyl nitrate* (*PAN*)]. These chemicals can act as potent oxidizing agents on all the cells in the respiratory epithelium. This reaction results in increased rates of aging processes and also increased rate in the amount of scar tissue formation in the lungs. Susceptible patients (those with asthma, bronchitis, and emphysema) will show increased symptoms of their disease at ambient levels of 0.2 parts per million or higher (Figure 53).

Other pollutants whose health effects have not been studied are peroxyacetyl nitrate, ammonia, aldehydes, olefins, and many of the other aromatics. We can surmise that, when there is a combination of two or more of any of these pollutants, their effects can be multiplied.[109] It has been proven that SO_2 and ozone will act synergistically and produce severe health effects greater than either individual component. The medical community can only conclude that, as the air pollution increases, our patients' symptoms and illness rates will increase. Again, we must emphasize that a hydrogen economy would eliminate the majority of these disease-causing pollutants (Figure 54).

Hydrogen vehicles are offered as the answer to the severe air pollution problem. Over 50% of present toxic pollutants in the South Coast Air Basin are produced from vehicular emissions. A scenario wherein the seven million vehicles of the South Coast Air Basin are converted to hydrogen would improve air quality sufficiently to fulfill the Clean Air Act requirements for 1985 (Figure 55).

4. Summary

The Riverside hydrogen bus does, indeed, represent an engineering technological breakthrough by utilizing the latest conversion devices for a clean nonfossil fuel vehicle. In addition to a historical engineering feat, it also represents a major advancement in fields of sociology, economics, and public health. Now that hydrogen has been demonstrated to be a viable alternative fuel, society can step from foreign fossil fuel economy into the clean "domestic" energy scenario.

Freedom from dependence on foreign oil, ability to establish our own national vehicular fuels, and potential for clean air while continuing our "single car convenience" will reprioritize federal budgets for energy, federal trade, pollution control, and public health. Universal hydrogen transportation will end the need for governmental monitoring, legislation, and policing efforts for maintaining decent air quality. Cars, trucks, and buses powered by hydrogen will no longer require the millions of dollars in pollution control efforts which have plagued our transportation industry for the last several decades.

Addendum to Paper by Robert M. Zweig

The medical community would like to offer an item for deep consideration in the deliberation on alternative fuels of the future: better health for us all and for our children—to the degree that it is affected by the polluting emissions of automotive engines. The substance of this paper is based on our experience in Southern California, and particularly in Riverside County.

California's South Coast Air Basin consists of four counties: Los Angeles, Orange, San Bernadino, and Riverside. In the U.S. and possibly the world, this area is among the most heavily impacted by photochemical oxidation—toxic pollutants that include the visible yellow haze of nitrogen dioxide, and invisible ozone, and peroxyacetyl nitrate—and by the particulate pollutants, the most serious of which are the hydrocarbon aerosols and the sulfate and nitrate acid aerosols. In comparing the stationary and mobile sources of these contaminants, we find that the greatest parts of carbon monoxide, the hydrocarbons, and the NO_x come from mobile sources; the SO_x comes mainly from stationary sources. Most of the conjecture at this Symposium and, generally, about alternate automotive fuels begins

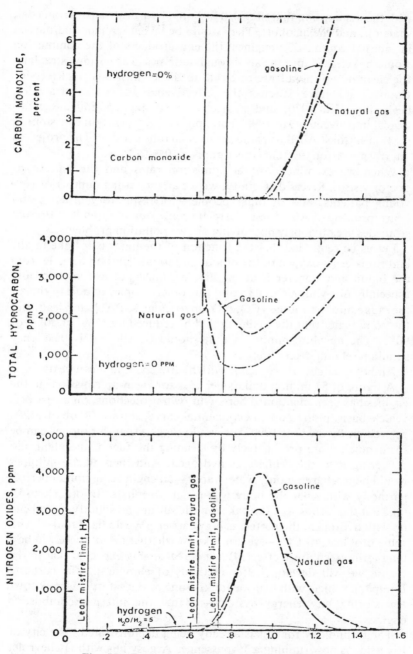

Fig. 55 Equivalence ratio (fuel lean/fuel rich)

with the assumption that they will be liquid hydrocarbons—from coal, shale oil, and the alcohols. They would be mixed with air and burned in internal combustion engines; the end-products of the combustion would be released into the air through tailpipes. The ensuing strategies for cleaning up the air are to clean up those emissions. In Riverside County, we have studied the effectiveness of these strategies, beginning with 1970, and projected, assuming continued technological improvement, to 1990. Total present hydrocarbon emissions in the South Coast Air Basin amount to 2000 tons per day. The projected improvement for 1990 cuts this figure by 10% to 15%.

When we examine hospital admission rates and the pattern of doctor vists in Riverside County, we see an increasing number of very young patients, and of older people with respiratory diseases and heart problems. And these correlate with our air quality. Present strategies are obviously not solving our air pollution problems.

Let us be reminded that the function of the lungs is to take in air, and transfer its oxygen to the blood; and to take carbon dioxide from the blood and transfer it ot the air. We must not ask anyone, and especially our young children and our older people, to add to the air they take into their lungs: CO, NO_x, particulates, PAN, and lead. But the point is not that the toxins should be reduced by 15%, or 50%, or 90%. The absolute minimum that should be allowed is 0%—they should be eliminated entirely.

And they can be. By refueling with hydrogen. And it can be done.

A figure of $1 million dollars per year for clean-up emissions in the engines has been mentioned here. Our own projections, on a cost-per-vehicle basis, plotted on an exponential curve, show a figure of $1200 per vehicle by the year 1990. We would suggest that this money can be spent much more productively by cleaning the fuel at the plant and delivering it to the vehicles as hydrogen. And then all the tailpipes would be discharging only water vapor—an engine emission in perfect harmony with what the body was meant to inhale. In other words, take out the carbon as CO_2; take out the sulfur and solidify it; and end up with hydrogen, the perfect energy carrier. (We do have to deal with some problems of CO_2—we don't know whether or not there will be a long-term greenhouse effect. But it is reasonably expected that in the future we will develop better methods of eliminating the carbon.) Equip the vehicle with hydride tanks—one of the safest devices known for storing any energy system—to fuel the internal combustion engines.

The technology for this is already being demonstrated. The city of Riverside is now running a 19-passenger Argosy bus with a Chevrolet engine that runs on hydrogen. In Provo, Utah, they too have a prototype bus running on hydrogen. The technology for releasing the hydrogen to the engine differs somewhat in the two buses—and no doubt there are other possibilities. But it works—and with higher efficiency than with gasoline. And most important, without poisoning the air.

Looking further into the future, to 2000 or to 2020, we can envision an economy where solar energy in its various forms provides the input for getting hydrogen from water; where the hydrogen is stored as a gas, or in hydrides, or cryogenically; and where hydrogen, as the perfect fuel, answers the needs of all users, mobile and stationary. And where air pollution from combustion is but a memory.

The writer implores those who will be deciding on the alternative fuels of the future to look at the many regions that are suffering from this blight, and to think very seriously about hydrogen.

E.   Commentary by Symposium Participants

(a) *Question:* Is there consumer support for any of these initiatives, and, to the degree that initial costs will be higher, can that support be maintained or enlarged?

Answer: There is some difference of opinion. The public in Nebraska appears willing to buy Gasohol even at 2 to 5 cents a gallon more than gasoline. However, they have an economic interest in promoting ethanol. When there is not this kind of motivation, you find instances where, for example, people put leaded gasoline into the "unleaded" tank, just because it is cheaper. The disagreement here is whether lower cost is indeed the primary reason. Might it not be first that their cars are running less well, and only second that it is cheaper? The typical Californian, for example, seems to support measures for cleaning the environment as long as he or she is convinced that they will work. Opinion polls bear that out. And even on the federal level, money is voted with the support of the public for programs of moral commitment. There is a fundamental dichotomy here, however: the public, when it is asked its opinion, or when it is constrained by laws, acts one way. But when it is asked to fork over its money, it acts quite differently. It is certainly questionable whether the success California has had with its vehicle regulations would be as complete if they were optional: if every automobile buyer were given his or her choice between an emission-controlled car and the kind they sell in the rest of the country. And there is yet another point. These issues do not just involve an idea on the one hand and the public on the other. There is a huge business establishment which forms a channel between the two, and regardless of whether the public is morally or economically motivated, there is no question that, for the channel, the motivation is always economic. Fuel distributors, for example, given the choice of product, will go to the one that is cheaper, regardless of the social good. That's business, and it is understandable; and the social good may or may not be served. However, one result in this instance is that (with the exception of the OPEC countries of course) this country has the cheapest gasoline in the world. And the public seems to be pleased about that. Another point is that the public is fickle. As the Proposition 13 vote in California has shown, when the public finds out how expensive the things it once wanted have become, it can change its collective mind. There are, however, other interpretations of the meaning of the vote on Proposition 13. One is that, rather than being a challenge to the idea of taxation generally, it is a reaction to a perception of not being well served—a feeling that "you're taking more and more of my money and giving me less and less in return." It is an expression of dissatisfaction, a vote *against*—which, political scientists tell us, is the way the public usually prefers to vote. And in respect to energy matters, we can sense an undercurrent of feeling in the community that, if another break in the flow of Arab oil can cripple our country, then the private sector of the oil business obviously can't "cut it," and had better get out, because there's too much at stake. It would be a terrible answer, but since legislators react to their constituents, it could happen. This assessment of public feeling is not necessarily everyone's. Another viewpoint is that the people, on the contrary, don't really believe that there is a problem. For example, when the public *did* believe it, there was great popular pressure to hold to the 55 mph limit, one driver on another. Now at 55 mph one gets blown off the road. Until the public really believes there's a problem, it will be very difficult to get their support for any costly alternatives.

(b) The Seattle operation might have made it had they received federal or state support, and instead of listening to a story of failure we would all be trying to duplicate their success. The government *does* subsidize the oil industry at about a billion dollars a month, and what we get for that, in say Southern California, is 2000 tons of fossil fuel pollutants per day. And that is paid for, one way or another, by somebody else. Perhaps if the oil companies had to do the paying, a cleaner product would be on the market. Or perhaps the government subsidy should be a *garbage* depletion allowance to clean up the fuels. In any case, something must be done by the government because people are dying—there is a day-to-day parallel in the South Coast Air Basin between pulmonary/coronary death rate and the smog condition. Either this "something" must be done at the refinery level or at the user level. But political figures have a very clear social obligation and cannot sit by and do nothing.

(c) *Question:* Since many of the pollutants come from smaller sources—motorcycles, lawn mowers, etc.—might we not get some quick results by promoting electric battery power for such uses? Perhaps in parallel with hydrogen?

Answer: Hydrogen and electricity working together are probably the ideal situation, but one would have to look very carefully at the source of the electricity before determining whether the tradeoff was beneficial. For example, in the South Coast Air Basin, the pollutants from the generators are becoming at least as much of a problem as those from internal combustion engines. This has become true since the shortage of natural-gas fuel stock for electricity has forced the use of high sulfur fuels. SO_x has now become a new and serious problem.

(d) *Question:* It is clear that there is no safety problem in storing the hydrogen in the vehicle. Using the hydrides is extremely safe. But what about in the use of the gas itself in the vehicle?

Answer: There is a very small amount of gaseous hydrogen in the pipeline from the hydride tank to the carburetor; this is the only part of the system that may not be faultless. In case of an accident it would probably ignite. However, because the amount is so small, it would burn out very, very quickly and not burn back into the tank.

(e) *Question:* What is the California licensing situation for putting a hydrogen vehicle on the road?

Answer: The California Transportation Committee and the Department of Motor Vehicles have granted a waiver for the use of the one bus. Should the time come when a number of hydrogen buses are in the offing, Riverside will "plug into" a state gaseous-fuel law that was passed some four or five years ago to encourage the use of propane (because of its low-emission qualities).

F. Workshop Session Discussion

It could be useful to look at some of the successful experiences dealing with impacts and institutional issues and see if we can locate the key ingredients that made them successful.

1. Consumer Response

One must be very careful when looking at the results of consumer surveys. There is a big difference between what people say and what they do. They say, for example, that they want stripped-down, fuel-economic cars, but that's not what they buy. And when they say they'll pay an extra 5 to 7 cents a gallon for an alternative fuel, that may hold while the idea is novel, but it certainly is not a response on which one bases an industry.

What consumers may be willing to pay could be beside the point, because they probably will not have so many choices in the future. The auto industry is going to spend some $80 billion in the next seven years, and will be committing to engines that meet social goals prescribed by law. That will have much more to do with which way we go with alternate fuels than will the preferences of the consumer.

It has been hypothesized that public behavior in the market-place—in this instance, how much the public will pay for fuel—is largely governed by legal constraints that come out of the political process, and that process, in turn, is largely governed by people's opinions as expressed in public opinion polls. These, however, reflect an inaccurate reality to the extent that people's behavior does not necessarily correlate with their words, as has just been pointed out. Thus the hypothesis appears to be incorrect, at least as it applies to states with diverse populations. Public opinion polls are not what influence legislators: it is usually competing interests. In the case of energy, what makes it especially difficult is that the competing in-terests are of equal strength. As for the public, since its behavior is emotional rather than logical in that it wants what it wants regardless of price, and will spend its money to get it (e.g., television sets, freedom of mobility and hence automobile transportation), this further complicates the job of the legislators. Such hackneyed issues as "big business versus the environment" turn out to be vast sim-plications, if not irrelevancies.

The often-referred-to American love affair with the automobile is not an explanation of the function of the automobile in this society. One might better speak of an American love affair with single-family housing at low densities, a pattern of land use necessitating some form of personal transportation. Transit cannot serve the configurations that have developed; a bus carrying two passengers is not only uneconomic but is wasting fuel. Since a half century or more is required for significant change in land use patterns, the personal vehicle is here to stay for the foreseeable future. A concern with transit, car and van pools, staggered work hours, flex time, and all the other reduction-oriented remedies is admirable, but the net reduction in gasoline use resulting from such often enormously costly efforts will be less than 5%.

Stated simplistically, the problem created by population preferences and technology can be dealt with only with substitute technology in the form of new fuels and engines. As a number of speakers have noted, putting such a system in place is a time-consuming process. If, however, the necessary technology substitution is not undertaken, severe disruptions of all aspects of U.S. society become highly probable (in about 25 years).

It might be noted in this context that, while the government is spending billions on energy-related R&D, the need for orderly technology substitution is nowhere explicitly recognized in such energy policy as it exists. Were the need recognized, the process of forming a supportive coalition would be helped greatly.

•

One positive aspect of the Gasohol idea (see further discussion below), and one which alone should cause the energy and oil industries to encourage and back it, is that it gets the public involved. "Gasohol" programs are implemented more or less locally, on a small scale. They get good press coverage. They get people thinking about Gasohol in terms of their own cars. They have a lot of environmentalist backing. And since individual farmers are involved, they get a good deal of grass-roots attention. All of this is in contrast to the apathetic public attitude toward the things the energy industry could (and will) do—economically viable, cost-effective processes—to come up with alternatives, because that is high-technology stuff, not understandable, and therefore uninteresting. In short, anything that maintains a high level of public involvement with the energy question deserves from the energy establishment all the real help it can get.

A large ingredient in the public interest in Gasohol is an apparent distrust of big oil and big coal. People have legitimate reason to complain about noxious air and strip-mine despoilage of land; they see alcohol as a step in the right direction. The California legislature is in part responding to this public feeling. Private interests, in their own self-interest, would be wise to align themselves genuinely and credibly with these public interests.

2. The Role of Fuel Cost

There are two issues here: fuel economy, and how much the auto makers are going to spend on achieving it, is one; which fuel we're going to put in the tank is the other. And that one includes looking at how much it's going to cost to make the fuel, because it's going to cost a great deal. Since that is one of the elements in the equation that will ultimately yield an alternate fuel, we should begin to deal with it and get used to dealing with it. One way to do that is to be arbitrary: pick a year—2000, say, or 2010—and pick a percentage—10%, 15%, whatever—which represents the alternative fuel portion of the total supply; then figure out how much it's going to cost to get there. It will be a lot, and we will have to get used to the idea that we're going to have to spend it.

With regard to costs, it can be useful to look at the failures as well as the success stories. The Seattle experience has certain elements in common with a short-lived foray that Exxon had made into the ammonia business. The lesson to be learned is this: in trying something new, one must be sure it is either a familiar area, or, if it is unfamiliar, that there is enough economic staying power to weather what almost certainly will be a series of initial disasters. This is the cost of learning. Staying power is what was missing in Seattle.

This capability is one that only the largest energy companies possess, and is an argument against divestiture. Energy companies, by definition, should be involved in all forms of energy, and so long as there is competition among equals, largeness here is an advantage. Only the giants can afford the risks inherent in alternative energy development. A rule of thumb is that, before a company embarks on a project with a 50% chance of failure, it should be able to afford to do five like that. Otherwise the risk is unacceptable.

A very sizable cost factor is one that is imposed by the federal bureaucracy: the cost of the time it takes for a new energy project to be approved or disapproved. What with the involvement of EPA, DOE, DOT, OSHA, FAA, etc., we get a long sequence of agency upon agency, each having to pass on the plans. This sequential procedure has been known to hold up the works for 7 to 12 years. In Colorado, on the state level, they have devised a system that effectively shortens this period to between 12 and 18 months: the company goes through all of the agencies in parallel rather than in sequence. It means more paperwork for the company, because the same information must be worked up in various formats for the various agencies, but the waiting time for decision is cut down significantly. The idea is being looked at in Washington.

3. Gasohol Incentives

The new California bill which gives the distributor a state-income-tax credit of one dollar per gallon of alcohol used in gasoline is very specific. The tax credit will not be given for any alternate fuel except alcohol derived from California agricultural products, and then only when used in a gasoline-alcohol blend. The blend can be in any proportion; there are no restrictions on that. The distributor will get his dollar-a-gallon credit against his California income tax for every gallon he buys, provided he can prove that he blended it with gasoline and put the blend into the gasoline distribution system. A provision calls for independent analysis after five years to determine how the law is working, and for suggested modifications. If no changes are recommended, all will continue for five more years. The idea is to provide a subsidy to allow anyone who would like to go in that direction to compete with shale, coal, etc., and thus help biomass get a foothold.

It is believed by some that the purpose of the California bill is not to solve the problem of how to get enough alternative fuel in use to keep us going while the oil grows short, but rather to serve the economic needs of farmers in the state. Otherwise the qualifications for subsidy would not be so narrow.

California's viewpoint is that this is sensible regional policy. Before the bill was brought out, there was consultation with oil companies, utility companies, the Chamber of Commerce, financial institutions, and labor organizations, and all were in favor of it. It is true that it is not purely an energy bill, nor is it intended to be. It is trying to go some way toward solving a number of problems. Energy is among them. So is farm waste material and farm surplus. So is unemployment. But we must remember that this is not proposed as *the* answer to the growing petroleum shortage. Other programs, federal programs to encourage production from coal and shale, for example, will go on. This is simply one route which is easily integrated into our energy system (witness Brazil), which at the same time helps some farmers and gets rid of a lot of waste that cannot be handled any other way, and should also, in its small way, help generally to beef up the state's economy.

Comparing the California incentive program for Gasohol with the working program in Brazil, we find that the Brazilians need no subsidies (and no supporting bureaucratic structure for administering) because the price of automobile fuel is twice as high as it is here. If we were allowed to do that here, we too would need no incentive programs. If the market price were a true market price, reflecting all costs including social costs, then the energy companies would assume their risks; and the higher cost of fuel would exact more rational behavior when the consumer chooses a car. He or she would insist on high fuel efficiency.

But to defend ethanol strictly on energy balance is to lose the case, because it is not quite the energy equivalent of gasoline. However, when other factors are included, the case for ethanol becomes persuasive. Some of the factors have to do with public involvement, as mentioned earlier. Another is that ethanol has the advantage of being ready to enter the market. Once it does, it could provide us with the experience and stimulus to work up a way of producing it economically. As for success stories, let us look at Germany: the reason that alcohol fuel is so much farther along there is that all of the parties involved—industry and government—are cooperating toward a national goal. Perhaps we can learn from them how not to be at each other's throats, how to change our thinking so that we can work together.

In Europe generally, and in Germany in this instance, problems of this kind are attacked rather differently: the government defines the problem and then goes to one of the largest oil companies, or one of the largest auto companies, or one of the largest engine companies, puts their best people together, and directs them to come up with an answer. That answer is respected by the public, and the work proceeds. If we were to try that here, the answer would *not* be respected. Big industry and government have a credibility problem here. Their conclusions are suspect even before they are announced. One of the reasons that the Gasohol idea is doing so well with the public is that it seems to have come up from the grass roots. The public has the sense that nobody is trying to rip them off with this one, that this is one alternative energy option that has nothing to do with big oil.

4. Government/Industry Interaction

If the key to Nebraska's and California's success stories is indeed getting the decision-makers in government together with the decision-makers in industry, then we have a strong message for Capitol Hill. But we must emphasize that the people who get together, on both sides, must be the right people, the ones who can see past the intermediate problems to the end solution. It is all too easy to get bogged down in trying to solve the problems along the way when one does not have a defined goal before starting.

The big question, therefore, is how do we achieve this? People from industry are motivated by the separate companies that pay them. Government officials are supposed to have a broader outlook, but still have their constituents, which include special interests lobbying for their own advantage. How do you get people to leave all that outside when they meet to deliberate?

That may be a question that doesn't need to be answered. Perhaps the answer takes care of itself when the right ingredients for success are present; and those ingredients seem to get included when the interest base is broadened. Because of social issues and international issues, alternative-fuel production cannot proceed in a free market (under which circumstance, it is generally agreed, there would be no need for this Symposium). But given a controlled market, and the problems we are faced with, the chances of success increase as more segments of society perceive a direct stake for themselves—when the farmers want it, when labor wants it, etc. In short, if technologists believe that we will settle the alternative-fuel question strictly on the basis of technical and economic feasibility, they are mistaken.

There is a related pitfall the technical community must avoid, and that is the tendency to be too insulated technologically—to proceed toward a solution within too narrow a model. In the past we have rushed ahead without really looking at safety, without really looking at the environment. Devil's advocates in these areas must be brought in from the very start. And in fact the environmentalist groups can be of help by joining in support of industry in asking for demonstration projects, because that is the only way to prove out the environmental impacts. This idea of cooperation is not only practical in that it attacks problems early, while they are still on paper, but in a subtler and even more important sense: it helps establish credibility. If all the participants in the push toward alternative fuels are perceived by each other and by the public as working in the general interests toward the same end, the obstacles to implementation will topple much more quickly.

G. Overviews

1. Peter Delescu

As a professor in the Department of Biology at the University of Santa Clara, I have, over a period of three or four years, been conducting research to determine specific environmental consequences of methanol spills in marine situations. I embarked on this research with the intention of exposing all the negative aspects of the fuel. But I have found very few.

Our early experiments were designed to compare spills of methanol with those of gasoline, crude oil, and ethanol. The results of these showed that the consequences of a major oceanic, coastal, or even fresh-water methanol spill were very temporary, except in cases so extreme that the concentration exceeds 1% in the environment, e.g., right at the site of a million-gallon spill. Research then proceeded in order to determine the physiological effects on marine organisms under different pollution circumstances: the effects at different concentrations of a single spill; and the long-term effects of a methanol leak in, say, a nearby pumping station. In addition, we want to understand the overall environmental situation in order to determine what action must be taken in the event of a spill. We are presently trying to develop a mathematical tool that will enable us to predict the consequences of a spill very rapidly.

We should be aware that methanol, as well as other alcohols, is normally produced in the environment, usually in very low concentrations because it evaporates quickly. But under not uncommon conditions, i.e., when certain films are formed over the surface of the water, concentrations of naturally produced methanol can exceed 1%. In coastal splash zones where water tends to stagnate, and in muddy environments, the concentrations go much higher—from 5 to 10%. Most coastal, inner-tidal-zone, and fresh-water organisms can tolerate significant amounts of methanol because they are continually exposed to it and are physiologically adapted for it. The crustaceans—crabs, lobsters, crayfish, barnacles, etc.—tolerate methanol much more readily than they do ethanol. This is the reverse of the response of the vertebrates (including man), for whom methanol is highly toxic and ethanol typically causes nothing more permanent than a hangover. There are certain crayfish, for example, that can live in as much as 10% methanol for five hours with no detrimental effect. Though this is an extreme, it serves to indicate that even in a major spill in a river system, if the water flow is large enough, there would be a minimal effect.

There are certain organisms that are very fragile and cannot tolerate methanol. These, however, are either very rare in the marine environment, so that their loss does not affect the ecosystem, or, like the ecoderms (starfish, sand dollars, and urchins, which are very important components of the seas), will recolonize so quickly that even a major spill would have little effect.

In addition to environment concerns, there is an economic concern: namely, what effect would methanol spills or leaks have on the commercial crab? The answer is that unless the concentration of methanol gets to about 5%, which is very unlikely since it is so miscible in sea water, the only effect on the crab would be a temporary "high."

The next question is what action is called for in case of a methanol spill? The answer would depend on the severity of the spill. Unless it were very large, or if there were a long-term heavy chronic leak, the best thing to do would be nothing. The disruption of the habitat during the cleanup would be worse than the consequences of the spill itself. However, if the spill were of such magnitude that a reaction were called for, then the most effective course would be to oxygenate the water. High levels of oxygen in the water enable marine invertebrates and vertebrates to tolerate much higher levels of alcohol pollutants. This can effectively be done by sailing a fireboat to the middle of the spill and spraying the surface of the water.

With regard to relative effects of methanol spills as compared with oil spills, it depends on where they happen. On the open sea, the consequences of a methanol spill would be greater. Oil spills at sea, so long as they don't come near the land, are very like a natural release of petroleum at sea from a subterranean fissure. Bacteria will ultimately break the oil down. However, in the case of a coastal or near-coastal spill, methanol is much less damaging than oil. Oil coats the feathers of birds so that they cannot maintain their body temperature, and coats the breathing and feeding apparatus of coastal marine organisms. Methanol, on the other hand, penetrates the water, but becomes very much diluted; it is very quickly biodegradable by marine bacteria and it evaporates quickly.

Should methanol get into a source of drinking water, and it was an underground source, the methanol would probably be metabolized by bacteria before it reached us. If it were in a small pond, and the polluted water were drunk right away, say by cattle, there could be some damage. But it would have to be drunk in very large quantities. All things considered, it would not seem to pose a real problem. There has, incidentally, also been research done on vertebrate inhalation of methanol. At saturation levels, the symptons resemble drunkenness—the basal metabolism goes way down. However, when the methanol is removed, recovery is complete. This then is an aspect that should be considered for anyone working in a methanol "gas station" where constant inhalation of low levels of methanol is possible. But the long-term effects apparently are minimal.

The conclusions we have come to are that, while methanol can be disruptive to the marine environment, it is a problem only under extreme conditions; and that compared with the alternatives it is a fine fuel.

2.   Edward M. Dickson

At SRI International—formerly Stanford Research Institute—we have performed a study for EPA and ERDA (now DOE) on the impacts of synthetic liquid fuel development. In the study, we identified a long list of factors potentially critical to attaining use of these fuels. I will describe some of these factors here.

We looked at three alternatives: syncrude from coal, syncrude from oil shale, and methanol from coal. We projected production sites for these fuels in various parts of the country, compared the obstacles to production at each of the sites, and evaluated the impacts. Here are some of the more pertinent findings:

(a) Methanol derived from coal is a significantly inferior option to syncrude derived from coal. This is true in all regions: the West, the Illinois region, or Appalachia.

(b) Comparing oil shale with coal as sources of syncrude, we find that oil shale, although closer to technological readiness, is a less attractive option than coal when all considerations are weighed.

In evaluating the options, we compared the amounts of the following inputs required to provide equal output of energy from each of the plants: capital, water, labor, and (for the coal options) coal. Breaking these down, we found that capital requirements are greater for methanol-from-coal than for syncrude-from-coal. Methanol requires about the same amount of water, but more labor and significantly more coal.

Although the syncrudes get higher marks than the methanol, *all* of the options face sizable obstacles which vary depending on the plant site. They all require large amounts of capital. Their water requirements become a problem in the arid West, where rich coal and shale deposits are found. They all require large labor forces which can create especially difficult problems in previously rural sections.

For example, in the arid West, where a typical large village is home to 5000 people, operation of a 100,000-barrel-per-day shale oil plant would take a community of 10,000 people. This can be very disruptive to the existing villages, creating boom-town conditions. (Congress is considering an Energy Impact Assistance bill to help communities deal with these problems should such plants be built.)

We also examined the issue of air quality and determined that plants for all three alternative fuels, using the best available control technology, would release approximately the same total amount of pollutants. The total pollution would be considerable. The composition of the emissions, however, would vary according to the technology. Any of the synfuel plants would emit more than twice as many contaminants as a well-controlled, very large refinery in operation today. (An interesting sidelight to this is that the total pollution from a synfuel plant, large as it would be, wouldn't even come close to matching the total pollution caused by the autos burning the fuel that the plant produced.)

An additional problem with methanol, besides the ones already mentioned, is that it would not enter the existing automotive fuel delivery system until the final distribution stage. In many cases the system would need modification to handle it. By contrast, the syncrudes would enter the fuel delivery system at the front end as just one more kind of crude oil. They would go through existing pipelines to the refineries (where any necessary small modifications would be centered); then through the existing distribution system to existing uses, most of which would require no modification. The present automotive fuel system is resilient, robust, and very expensive, and there's a lot of motivation to keep it working just as it is. Adding one more source of crude oil would seem very natural to the participants in the system.

Another issue that we examined in our study was the question of resource depletion. Since we are seeking alternative fuels precisely because we are entering an era where our present fuel will be difficult to obtain, we should certainly look at the resource base for proposed alternatives to determine how long they will last. We should seek to make the supply last as long as possible. Our most plentiful resource is coal. When we took into account the net energy ratio of the leading contenders (that is, the total energy in the original resource compared to the energy spent in getting it into the proper form and delivered to the market) we found that syncrude from coal was the most conservative way to use our most plentiful resource.

To project how long the coal would last in a coal-based synthetic-fuel economy, we constructed some plausible scenarios of energy use to the year 2050. We found that by 2050 we would have used one third of our coal and committed another sixth to be used in then-existing plants for their lifetimes. And the "resources remaining" curve would be dropping so fast that it would hit zero—no coal—by 2100.

We then broadened our analysis to include additional coal conversion possibilities: Fischer-Tropsch synthetic gasoline, hydrogen from coal, methanol, methane, and electricity. We found that Fischer-Tropsch would use nearly three times the coal of the syncrude option. In this analysis, our net energy ratio extended up to the point of energy delivered to a vehicle. When we extended it further to the drive wheels of a car, and included assumed engine efficiencies in the calculation, we found that the most energy-effective total system for alternative propulsion was the electric automobile, followed by the syncrude gasoline auto, the methanol and liquid methane autos, the hydrogen auto, and, finally, the Fischer-Tropsch gasoline-fueled auto. When we calculated energy cost, delivered to the vehicle, we found that, following the present natural-crude gasoline, the cheapest fuel would be syncrude, followed in order by methanol, liquid methane, liquid hydrogen, and, finally, Fischer-Tropsch gasoline.

We found that the coal cost was not the dominant factor; instead, the economic rule of thumb is that, for all synthetic fuels derived from coal, two thirds of the price of the final product is a reflection of the cost of the capital required to build the plant, and one third reflects all the other costs—labor, coal, other raw materials, etc. And another rule of thumb for these very capital-intensive projects emerges: the higher the cost of an alternative, the more wasteful it is likely to be of our natural resources.

3. Michael Farmer

Many of the comments and some of the frustrations at this Symposium can perhaps be put in perspective with some quotes and observations of other people under other circumstances. To wit:

At a conference in November 1977 on initiatives for coal utilization, Dr. Monte Canfield, Director of the Energy and Mineral Division of the General Accounting Office (GAO), said "I think it's time we woke up and recognized (that) this country is crawling with all kinds of different social values, and energy development isn't the only one. I'm not prepared to sit back and say we'll trade away these kinds of values." Participants in this Symposium represent a variety of social values, as distinct from the technical expertise they bring with them, and this accounts, quite naturally, for some of the frustration that we are bound to feel under such circumstances.

Professor Donald Michael, in an article entitled "Technology Assessment in an Emerging World," lists three major pressures for assessing technology: (a) A growing tendency to re-evaluate the priority of science and technology as a social enterprise. There is an accumulating perception that science and technology are not objective and disinterested; that choices of research topics and positions are expressions of subjective interests. (b) A growing appreciation that societal survival requires a systems perspective, an ecological perspective, a holistic perspective. A bits-and-pieces approach will not do. (c) An increasing demand by citizens for participation in decisions affecting their destiny.

All of this is relevant to the present discussion because most of what appears in these Proceedings is technological, and in a technological frame of reference. This prevents us from reaching agreement on many fundamental issues. In contrast to West Germany, Brazil, Japan, and South Africa, where alternative fuel programs are progressing rather smoothly, we in the U.S. represent a very large number of polarized priorities out of which it is difficult to achieve a consensus. It is reflected in this Symposium by the variety of implicit assumptions which conflict because they have not been arrived at on any common basis. A common basis is acknowledgedly difficult to find. But some help in organizing the logic of our assumptions may be forthcoming in work done by Clive Simmonds of Canada's National Research Council.

Mr. Simmonds said that the way in which different industries compete provides a basis for industry classification and a basis for predicting their future behavior. He broke down industries (including government) into eleven classifications:

(1)   Material resource—agriculture, forestry, mining, etc.

(2) Cost-minimizing—petroleum,   energy   products,   transportation, utilities, etc.

(3) Performance-maximizing—chemicals,   computers,   electrical devices, etc.

(4)   Sales-maximizing—consumer products.

(5)   Service—personal, community, and business (including entertainment, etc.).

(6)   Social   betterment—education,   environment,   health,   and welfare.

(7)   Capital goods.

(8)   Distribution—wholesale, trade, imports, and exports.

(9)   Finance.

(10)   Security   and   protection—police,   fire   brigades,   the   Armed Forces.

(11)   Administration and management, including federal, state, and local governments, and regulatory agencies.

An awareness of the different motivations that separate these industries can help us to understand the assumptions they bring with them. That sort of understanding might help us order our own, and get closer to a consensus than we seem to have been able to do so far.

4. Peter Meier

A look at the history of the solid-waste management and resource-recovery field will raise some questions that might profitably be asked in the alternative-fuel field. The questions are: (a) What are the relative merits of high technology versus low technology, especially in regard to timing? (b) What is the role of government, and state government in particular, in creating a market? (c) What is the role of government, and state government in particular, in assuming some of the risks?

The late 60's and early 70's saw the beginning of a number of high-technology approaches to resource recovery from refuse. Monsanto had a pyrolysis process. Union Carbide had a pyrolysis process. Union Electric in St. Louis, going on the assumption that the best use of refuse is to burn it, set up a large-scale demonstration project using refuse as a supplementary fuel for their boilers.

Let us look at what has happened. Monsanto, with a promise of steam as its ultimate product, found a customer, signed contracts, and thus raised the financing to build a pyrolysis plant in Baltimore. The only problem is that the huge Monsanto complex mostly doesn't work—and when it does, it doesn't work very well. Union Carbide, for its part, got a very enthusiastic response from Westchester County in New York. But as the years went on, the capital costs kept escalating. Westchester has decided to go to a conventional land-fill. And Union Electric, when it wanted to move from a demonstration project to a full-scale process, met a brand new set of problems —raising finances and finding sites for transfer stations where refuse would be transferred from collection trucks to the tractor trailers that would take it to the boilers.

The failure of the vaunted high-technology approaches has given the resource-recovery industry a bad name, and as a result, even the low-technology approaches are suffering. As a case in point, a modest low-technology resource-recovery plant is working in Baltimore County, but only because the county and the Maryland Environmental Service were willing to put up the money and assume the risk. Otherwise the plant would never have been built. Customers for the refuse-derived fuel, because of promises unkept by the high-technology people, were wary of entering into binding contracts until the plant was built, operating, and had proven itself. Without these contracts, normal financing could not be raised. So Maryland financed the plant and has taken on the risk of finding the customers, which should not now be too onerous, with the plant working well.

The lessons are clear. When the public interest is at stake, and there is a dearth of venture capital, there *is* a role for the state government

to play: to create an initial market and assume the initial risk. And by
beginning with low technology, the risk is less; but the ice is broken.
Extending this lesson to alternative fuels, the Nebraska Gasohol
example shows that, when a state is playing for low-technology stakes,
nothing succeeds like success. Iowa is about to follow Nebraska's
example. Illinois has already begun. It is a good bet that, within the
next few years, most of the farm belt states will be on the bandwagon.

5. Liquid Fuel: What Next—Alcohol?

Scott Sklar
New York State Alliance to Save Energy, Inc.
New York, New York

Abstract. Alcohol fuels offer promising potential for transportation
and industrial application. Alcohol, either ethanol or methanol, can
by synthesized from a variety of plentiful domestic resources including
urban waste, timber, agricultural waste and surpluses, coal, and even
algae. Alcohol can be immediately integrated into our present energy
system—through blending with gasoline as an automotive fuel, as a
pure fuel in test fleets, as an industrial turbine and utility peak-turbine
fuel, and for use as a base chemical for plastics and synthetic fabrics.
The multiplicity of end-use possibilities as well as its nonpolluting
characteristics make alcohol a valuable bionass fuel. Because of
alcohol's interchangeable role as a fuel or industrial chemical, it
appears more economically viable than many other potential liquid
synfuels. Additionally, one of the only ways to produce an environmentally acceptable fuel from coal may be through its synthesis
into alcohol. Because of the varied resource base and its possible
implementation in a centralized or decentralized system, alcohol can
be adapted to the needs of most geographical regions. The United
States has very few near-term liquid fuel options. Clearly shale oil and
other liquid synfuels cannot be commercialized in the immediate
future. U.S. balance of payments and the decline in the value of the
dollar will not be decelerated unless foreign petroleum purchases can
be offset. Time is running out for the U.S., other industrialized
powers, and the third-world nations. Alcohol fuels are the short-term
answer to the liquid fuel shortage because of their many end uses,
broad resource base, environmental attributes, and social benefits.
The time certainly has come for alcohol fuels.

Social and Political Analysis. The President has described the
energy crisis as "a moral equivalent of war." The Congress has
argued and delayed about regulation and deregulation. But no one has
addressed one of the most important energy issues—what will be our
next liquid fuel?

Solar technology will be the best answer to our space heating problems and will also meet some industrial needs. Geothermal and windpower, too, have their places; but none of these options can be harnessed for personalized transportation. There have been no successful solar or wind powered cars to date and I venture to predict that we will not see any in the near future.

The electric car has made some advances during the energy crunch. Now these cars may attain speeds of 45 mph for 100-mile intervals. Then their batteries must be recharged for a day. Nevertheless, the commercial practicability of the car is illustrated by Sears, Roebuck and Company, which plans to market an electric car shortly. The electric car, at present, is perfect for inner city travel but will not adequately meet the needs for the average driver.

What options remain? One is alcohol, which can be produced from a variety of sources—grain, sugar, agricultural wastes, garbage, timber and timber wastes, and coal. The use of alcohol as a transportation fuel has been around a long time: our earliest cars used alcohols as fuels, as do our race-cars today. During the last two years of World War II, Germany ran a significant part of its war machine off of alcohol. Both Germany and Brazil today are incorporating alcohol fuels into their economies. Brazil is relying on agricultural sources, mainly the manioc root which grows in great abundance and sugar cane and sugar beets which are currently being grown in energy farms.

The United States, so far, has failed to embrace alcohol fuels as a visable energy option. The three major reasons for this caution are:

(a) Concerns over adequacy of supply.

(b) Doubts about energy efficiency.

(c) Arguments concerning economics.

Supply is the least problematic issue. Methanol can be synthesized from coal, our most abundant and traditional energy source. Although methanol is not the best alcohol for blends with gasoline for motor fuel, it can be used alone as a most efficient pure motor fuel and utility-turbine fuel. Additionally, methanol can be derived from timber and urban waste as well as coal.

If we consider the facts that urban areas are running out of landfill sites, timber wastes are burned on-site in our forests, and agricultural wastes are poorly utilized, alcohol fuels may be able to play a part in the solution of other pressing national problems. The urban waste problem has deteriorated to the point where Philadelphia has been granted a permit to dump its garbage in the Atlantic Ocean. If the United States was able to convert 80% of its existing waste materials (garbage, timber wastes, agricultural wastes, manure, sewage, etc.), we would be able to meet 75% of our imported petroleum needs through alcohol or methane fuels.

The farmers are pressing Congress and the Department of Agriculture for alternative income sources for their products. Because of hugh surpluses, market fluctuations, contaminated foods, infected crops, droughts, and spoilage, much of what we grow is wasted. This waste has never been adequately used and could provide the farmer with another marketable commodity. The multibillion dollar farm bill passed by Congress bails out the farmers temporarily, but provides no long-term solutions—alternate income sources. In addition, new technologies give farmers the capability to harvest high-yield crops which would not be used for food. The best solution appears to allow farmers to plant energy farms on current land set-asides not utilized for current food production under the direction of the Department of Agriculture.

A major agricultural concern is what is the best way to dispose of contaminated grains and other crops. At present, corn infected with a carcinogenic mold this past year was sold on the black market for poultry and cattlefeed, burned on-site, or left in piles where the mold toxin infected the ground water. Alternate positive ways to dispose of these crops should be instituted. The Southwest Alabama Farm Cooperative is attempting to turn this infected grain into ethanol.

Edward Lipinsky, a Batelle Corporation researcher, had developed an interesting theory for a more efficient use of grains. Most of our grain is used for animal feeds and most of this animal feed grain goes to cattle. This cattle feed grain can be distilled into alcohol. The stillage (the sludge left over after the distilling process) is higher in protein content than the original grain. The stillage can then be mixed with the grain stalk (the stover), transforming it into a far better cattle feed than the original product. Not only do we have a quality cattle feed buy a fuel product as well. The resulting ethanol is an excellent motor fuel and can be blended with gasoline. If this method were widely employed we could have stillage available for other purposes than just cattle feed. Because this stillage is so high in protein, it could be used as a powder protein food supplement for our processed foods or exported to the poorer countries. The Domestic Technology Institute in Colorado has plans to produce ethanol, an animal feed, a protein food supplement, and an organic fertilizer in a similar project.

New technologies are surfacing which may revolutionize alcohol production. A process developed by Dr. Donald Brewer uses algae grown in waste water on waste land to produce ethanol. The process yields greater quantities of ethanol than nay other known process. Dr. Brewer is currently setting up a demonstration plant in Florida. The possibilities for using this new technology for energy farming are manifold because the investment costs are minimal.

Most people become easily confused with energy terms, particularly concerning "energy efficiency." It is described by experts as the energy consumed to produce the energy product. Obviously, it is theorized, if we use 100 Btu to produce 70 Btu we are not energy efficient. Unfortunately such a simplistic measurement may not adequately address the issue. Hypothetically, if we used a gallon of petroleum to prudce a gallon of methanol from garbage, our energy theorists would screem "energy inefficienty!" But they overlook the fact that we may be wasting several gallons of petroleum to dispose of this garbage elsewhere. Thus it is important for you not to become scared by energy code words. Thus it is important for you not to become scared by energy code words. We must be careful to look at the entire social picture and not be narrowly directed into only energy concerns.

I have studied many of the energy critiques opposing ethanol production by our nation's farmers. High on their list are concerns for energy efficiency. The critics assume that the fermentation times cannot be lessened and the plant base fuels will be petroleum fuels. This contention is absurd. Alcohol plants will most likely rely on cogenerated steam from utilities and other industries, straw, bagasse, corn stalks, solar energy, and even geothermal. Additionally, many energy policy makers overlook the fact that it takes three Btu of coal to produce one Btu of electricity in the home. I have heard no one suggest we stop using electricity as an energy source. Thus, the negative assertions concerning energy efficiency have falsely clouded the entire energy farming issue.

Finally I reach the fundamental issue for those who are cautious about alcohol fuels.

Economics. As previously stated, our early cars ran off of alcohol. Why, if alcohol is such a fine fuel, did we change to petroleum-based fuels? The answer parallels the reason why the United States stopped synthesizing gas before World War II—it was easier to drill a hole and let the already made fuel pour forth than make a fuel from scratch. Not only is it easier but less expensive as well.

Economics has always been the prime consideration behind the delay in implementing an alcohol fuel policy. This delay, caused by economic concerns, may be justified. However, the economic reasons have not been fully analyzed, nor have reliable cost projections been available. The following are the reasons for this lack of information and miscalculations on the economic viability of alcohol fuels:

(a) Petroleum price projections have been sketchy at best. No one really knows the relationship of cost to supply to the difficulty of extraction in the future. Thus accurate cost comparisons of alcohol fuels with gasoline are virtually impossible.

(b) Every new fuel option will require enormous capital investment. If we single out any new fuel technology, the costs, at first, appear prohibitive. We must realize that broad implementation of new technology will cost much in the short run, but this should not deter us from an alternative fuel policy.

(c) The initial costs of alcohol technologies have not been separated out from the standard market costs. It would not be fair to compare alcohol costs vis-a-vis gasoline at this time because we are comparing an infant industry in the early growth and development stage with an older and proven petroleum industry (i.e., the petroleum industry has received at least $2 billion per year from 1962 through 1972 in oil depletion allowances).

(d) The side benefit costs must be calculated in with the estimates of economic viability. If alcohol produced from garbage would save us the expense of waste processing, landfill acquisition, and disposal, then in "overall" terms garbage conversion to alcohol may be the most long-term economical approach from a total systems point of view. Also, benefits from an alcohol fuel industry should be addressed in terms of increased domestic employment, reduced pollution, balance of payments, and in displacement of farm subsidy costs.

(e) Raw materials for alcohol production vary from region to region. The cost disparity between ethanol from farm produce and methanol from coal is large. However, dependence on several domestic sources for our energy appears justified rather than reliance on seemingly inexpensive and finite resources which make us vulnerable to outside powers. We must prioritize the need for energy self-sufficiency versus a need for less expensive fuels.

(f) There is the universal problem of melding new alternative energy technologies with older existing technologies. Costs are usually jacked up because we must add the costs of converting already existing plants to accommodate the infusion of new technology. An example of this would be the added costs tied to the shipping of ethanol for gasoline blends to petroleum refineries which grew in areas most accessable to petroleum shipments. Obviously, large refineries and storage facilities could not be easily moved close to alcohol production sites.

In addition to the above six points, technological lag is another major obstacle to alcohol fuel implementation. Small decentralized stills were operating in this country during prohibition, while larger plants were in operation during World War II producing alcohol for jet fighter fuel additivies and as a base chemical for synthetic rubber. Modern alcohol production has relied solely on petroleum and natural gas to derive alcohols for the production of industrial chemicals.

Commercial ethanol plants exist today only for our domestic liquor industry.

In the last five years, numerous new technologies have been discovered, offering promising potential for significantly lowering the cost of alcohol. For example, Dr. George Tsao has made a major breakthrough with his 100% cellulose conversion process. Dr. Tsao, from Purdue University, claims the cost of alcohol (ethanol or methanol from biomass) could be brought down to at least 80¢ per gallon. Dr. Donald Brewer, mentioned earlier in this work in connection with his Algahol process, claims he can undercut the per gallon cost of Tsao's process.

Also illustrative of recent technologies is the production of *ethanol* from urban waste. This process, developed by The Solar Fuel Company of Manketo, Minnesota, is significant because ethanol has a higher Btu content than methanol and also has a greater proclivity for blending with gasoline. The flexibility of using urban waste to produce either ethanol or methanol makes it a most attractive fuel source.

Finally, one related technology that's worth mentioning is Methacoal. The process involves producing methanol from the coal and slurrying coal in the methanol. This slurry process does not have the large water requirements of conventional water slurry and the slurry is a cleaner fuel.

These four technological innovations are only illustrative of the many recent achievements in alcohol technology, and the future is sure to hold others. Clearly if these processes can lower costs substantially and use resources which would otherwise go unused, then our problems of economics and supply could be resolved.

Conclusion. There have been many unanswered questions regarding the implementation of alcohol technology. I hope I have addressed some of these concerns. Many people from the Department of Energy have privately expressed reservations concerning alcohol fuels because there may be in our energy future a more suitable end-product fuel from the same source materials used for the production of alcohol. Methane, synthetic gas, shale oil, biomass pellets, kerosene, etc., could also be developed as major fuel options from alcohol fuel sources. However, this uncertainty is no excuse for delay in implementing projects across the country to test the viability of alcohol fuels. There is no better way to test the technology and markets than by developing demonstration projects across the nation in order to make an informed decision as to the possibilities of this new fuel option.

Of course, we should remain flexible in our energy policy, but delay for delay's sake will never bring the United States closer to a partial fuel self-sufficiency.

The bibliography which appears in Appendix B was extracted from my 800-item annotated bibliography, which is available to all interested parties.

Addendum to Paper by Scott Sklar

The main problem illuminated by this Symposium is that we have too many things getting in the way of energy. We have environment to worry about, we have big government bureaucracies, the state and federal agencies, and regulatory matters. We have OSHA requirements, we have public-acceptance problems, and of course we have the politicians. We also have "mind-sets": e.g., "We were regulated 33 years ago and if Congress hadn't done that, things would be okay today"; or, "I'm a nice guy, I do all the right things—and as that old song in the 50's goes, 'Why is Everybody Always Picking on Me?' " And lastly we have ads in the newspapers by companies dealing with these issues telling their consumers "Congress is wrong; they don't know what they're talking about. We do."

A second major problem is credibility. When a company says, for example, "We're good guys. What's the problem?," a solution would be for the company to get its friends, go to the people, and *really* say that, so they can build up a kind of credibility. The credibility problem isn't just limited to the oil industry. And although there are individual people in industry who are able objectively to see the problems, understand the social implications, and understand the hard technology, it takes more than just one person to solve the problem, and it takes some kind of collaboration.

A third problem is illustrated by the (fictional) Acme Gas Company, which goes to Congress with its lobbyists and says, "We don't care about social ills. Social ills are not our bag. We don't understand them, and we don't care about them. We're here to provide people with gas, that's all."

A possible solution to this problem is the coalition approach—a kind of approach I advocate very strongly, but which is seldom used. It is best illustrated by a conversation I had with a friend of mine who's a lobbyist for a major oil company. He came to me one day and said, "Hey, why are you guys in Congress interested in alcohol fuels? Obviously you know that they can't do that much, that they can't supply all the need for liquid fuels, that they're a problem." I agreed with everything he said for an entire hour, but I responded by saying, "In Congress, on the alcohol-fuels issue I am able to get state governments, farm states, coal states, timber states, urban states, the minorities, the environmentalists, certain petroleum companies, and certain auto companies to work with us—on an energy issue, an immediate energy issue. No one else has done that with any other issues.'' My friend walked out with an understanding of this issue as a fine example of one that has built a broad coalition—a kind of public acceptance between people in the private sector, people in the government, people in the local regions, and the person on the street. This kind of coalition can be used in many other technologies.

We can also employ what I call a systems approach, which is primarily education. It you were to go on the streets of Santa Clara and talk about shale oil, I don't think there'd be anyone you'd meet who had the foggiest idea of what you were talking about, other than that you can get something out of shale, and the oil companies are going to manage it, and maybe it will get to them. Because there has been no "coalition" politics or "marketing" effort, there has been no substantive approach to the people on some of the real issues.

We can also undertake what I call "aggressive nonconventional liaison." There are certain technologies, certain ideas, that the private sector has had that have great social implication; and there has been a constant fear of bringing them out into the public arena. A perfect example is the chemical company that came to me and said, "We're thinking about making ethylene from ethanol from corn cobs." They went through an hour of all its great social implications. They even thought the economics would work out on its own. But the problem was that they were afraid to come out and open the idea up to public scrutiny and build up a public constituency. They claimed that "wasn't their job." I claim that it *is,* and that they'll never implement their idea until they recognize that.

Finally there's the mind-set that I call "put in just what I have to"; i.e., if I'm applying for government regulation, and am required to fill out the forms, I do so. That's all—that and no more. But it has been my experience that if there *is* personal contact, if there *is* a broad education effort made, a lot of regulatory hassles can be made much easier. The problem for the regulators is that they don't know, in many cases, some of the new technologies that they're regulating. And then, by surprise one day, something eight inches thick falls on their desk.

There's no question that government will have to reconcile what is good for the public with what is good for the private sector. That's not an easy task.

We also need some strong assertive government controls on imports. The only way that will be accomplished is to set up a coalition of private-sector and public-sector people in the country to agree upon a policy which limits imports, and then set up a work-plan to make up the difference.

Furthermore, there will be no advance in new energy technology without linking it to a consideration of social ills. The fact is that no one in Congress or on the state level can let anything happen which will adversely impact a large group of people who cannot help themselves. There must be an awareness of this need on the part of both industry and the public sector. It has been accomodated in the past, and it can be again.

It is also necessary to do what I call "dividing the kitty." There is no way that we can continue with the idea that, say, shale oil has to take the entire brunt, or that coal liquefaction is the only answer. Every viable technology has to have a part of the pie; that's how you get public support for new technologies. If someone feels, "Hey, this technology will cut me out," you won't get cooperation. This is the essence of coalition policy making.

Finally, it's necessary that we have immediate experimental development. There needs to be cooperation between the public and private sectors to get some demonstration projects on the road, in all the viable technologies. Such demonstrations are essential to generate the hard data that are needed so critically, and that we now have only paper statistics for.

6.　*Symposium Participants*

(a)　*Question:* What can we do to help make any of those suggestions work?

Answer: Stop thinking along adversary lines, which has long been a problem in energy development. The cross-current of information must flow not just between policy-makers and technologists, but among policy-makers, technologists, *and the public*. And it must be a genuine *cross*-current. The needs and concerns of the public must be understood, genuinely considered, and included in the solutions. Otherwise we're right back to the adversary system, and it doesn't work.

(b)　Lest we become overwhelmed by the apparent success of the Brazilian ethanol program, let us look at some of the ingredients of that success. To begin with, Brazil is more dependent upon imported oil than we are, as a percentage of their total energy supply, and this is creating a very significant balance-of-payments problem for them. Next, although they are now producing ethanol from sugar cane and sugar cane wastes, they have large potential crops for ethanol from certain nuts and roots that grow in the vast unsettled areas of the country. Given these motivations, they are willing to produce ethanol at twice the cost of imported gasoline (on a volume basis, or four times the cost on an energy basis). And yet they *sell* ethanol for less than gasoline. How do they do it? By cross-subsidy. The oil plants are government-owned and the Brazilian government can mandate whatever it pleases because the generals have been running the country for a long time now. To get that kind of alcohol fuel "success" here would require a similar ability to make things happen by fiat—sometimes antithetical to our form of government. Even to suggest a cross-subsidy of that kind here would be to court defeat at the polls, especially now, in the Proposition-13 climate in which the public wants a strong say as to how its money is used. In the light of this climate, if we look at the Nebraska Gasohol project—with its built-in enormous cross-subsidy—we can reasonably predict that it will have tough sledding.

The response to that is that the public supports programs when it believes a need is genuine. There is still no national consensus on the energy issue. The public is not about to support spending the kind of money needed to get sixty-five plants going for a coal liquefaction industry when the only ones selling the idea are such low-credibility groups as the energy companies and the government. Gasohol, on the other hand, has support because the public believes it will contribute to a healthier agriculture, a healthier employment situation, and, as a way of disposing of waste products, a cleaner environment—all of which are perceived as genuine needs.

(c)   For those who are impatient with the inability of Congress to assume leadership in alternative-fuel development, it should be re-emphasized that Congress is a reactive institution. What it is presently reacting to is many fragmented groups of special energy interests, each distrustful of all the others. The Congress will be able to react positively when the distrust is replaced by a realization that there is a role for everyone in solving the energy problems, and a viable consensus solution is arrived at. The technologist, on the other hand, is faced with the problem that his training and his interests are in his technology, not in the politicizing of his technology. Here he is too often a fish out of water, and the strain of having to be an interface between technology and political demands is something that he copes with poorly, and often at great personal sacrifice. But technology does not exist in a vacuum; it does serve the public. Thus there is a natural conflict, which has appeared even in this Symposium, between advocates of technology and advocates of institutional factors as to which one should be the primary concern. In any case there are certain fundamentals that both camps can agree on. Once they are acknowledged, we will be able to see more clearly how to proceed. For example, we know that, for highway transportation, the only sufficient domestic resources currently available are coal and oil shale. We know too that we can extend there resources by the use of renewable resources, that is, biomass. But since we still need real data, and those can only come from working demonstration plants, we must also marshal the necessary forces to proceed with intelligent and beneficial work on coal and oil shale. If we can agree on that, let us now grapple with the question of how to marshal these forces. The answer again seems to come back to credibility with the public. The technological, social, and industrial ingredients for setting up these plants are on such a scale that credibility will be hard to win. A lot of land would have to be moved—and people already are living next to coal slag heaps. A lot of people would be brought in, and a lot of others may be displaced. That too is an unpleasing prospect. And the only data to support these vast projects currently come from major companies and the government, where the trust quotient is low. One possible way to proceed is the way of the World War II productivity councils. These were regional groups of labor people, consumer people, industry people, attorneys, and local politicians who met, under the pressure of winning a war, to hammer out issues that affected them. If we can get similar regional groups organized to win the energy war, and make sure that all interested parties are represented, and really work together, the public would perceive the legitimacy of the work and would support it.

(d) *Question*: If the public does not really believe there is an energy crisis, how did the automobile fuel-economy laws come to get passed? Where did the widespread support for that expensive measure come from?

Answer: First, the laws were passed, not because of widespread support, but because of a lack of widespread opposition. Most of society did not feel strongly threatened by these laws. Second, individuals perceived the fuel-economy target as the automobile makers. They felt that the government was forcing the big auto companies to improve their product for the good of the individual citizen. That was a rather satisfying feeling. But individuals do not perceive that kind of benefit in alternative fuel legislation. So perhaps we could overcome public fears of new syncrude plants by addressing those fears. Instead of yelling "crisis"—which the public doesn't believe—we perhaps should approach these plants as a national insurance policy against a *future* crisis; and at the same time show the public that we are listening to them—that we are listening to their concerns about boom-towns and foul air and ash heaps, and that we will give those concerns as much attention as we do the physical plants themselves. And mean it. One approach might be to get experts together from the various energy sectors—hydrogen, coal, shale, biomass, solar—and have them agree on a national program of demonstration projects, one or two for each technology. Congress would probably accept it. For one thing, they would for the first time have an energy constituency that is in agreement rather than fragmented; a congressman could then vote "aye" without alienating a constituency. For another, among the different regions of the country we could accommodate the different technologies acceptably. Getting the experts to work together would not be easy, but it could be done *now*. We could get something on line to begin building by 1980, and with Congressional authorization. In real terms, that is probably the only way it will happen at the federal level, short of a real crisis. Otherwise, it will have to happen state by state by state. And it is not likely to happen state by state. The only things the states can encourage are easy things like putting grain alcohol into gasoline. They cannot handle the large-scale demonstrations we need to tell us exactly what the problems are—the institutional problems, the environmental problems—to get, say, syncrude out of shale oil. One of the reasons we can't seem to get demonstration projects accepted is because of the way it has been done in the past; i.e., the government gave a loan guarantee to a company and told it to go out and do it. There is no constituency for that; it just appears as if somebody is getting rich. One possibility is to make it cooperative, that is, get *several* companies to share the costs of a model demonstration. Furthermore, bring in the state, the country, the model-community builders, the environmentalists. Call it an experimental program, because that's what it is. Everyone would be watching. The more people there are that are involved, the more there would be rooting for it. The crisis may not be here now, but it is coming. We need the experimental programs now so that we will be ready to deal with it.

REFERENCES

[1]Shonka, D.B., et al., "Transportation Energy Conservation Data Book," 2nd ed., Oak Ridge National Lab., ORNL-5320, Oct. 1977, p. 161.

[2]Loebel, A.S., et al., "Transportation Energy Conservation Data Book, 1st ed., Oak Ridge National Lab., ORNL-5198 Special, Oct. 1976, pp. 63, 74; also see Ref. 1, pp. 164, 167.

[3]"National Undertakings to Improve the Efficiency of End Use of Energy in the United States," TRW Systems and Energy, April 15, 1977.

[4]"Market Oriented Program Planning Study (MOPPS), Vol. 3: Transportation Sector," U.S. Dept. of Energy, Assistant Secretary for Energy Technology, Final Rept., Review Draft, DOE/ET-0010/3(D), Dec. 1977.

[5]Balasubramaniam, M. and Jenkins, R., "Automotive Fuel Requirements with Increasing Diesel Car Population and Implications to Refining Operations," TRW Energy Systems Planning Div., prepared for U.S. Dept. of Energy, Office of Policy and Evaluation, Contract EX-76-C-10-3885, Feb. 1978.

[6]"Market-Oriented Program Planning Study (MOPPS), Vol. 1: Integrated Summary," U.S. Dept of Energy, Office of Energy Technology, Final Rept., Review Draft, Dec. 1977.

[7]"Analysis of Alternate Energy Forecasts," Brookhaven National Lab., Feb. 4, 1977.

[8]"Gillis, J.C., et al., "The Technical and Economic Feasibility of Some Alternative Fuels for Automotive Transportation," *IECED '75 Record*, p. 856.

[9]"The National Energy Plan," Executive Office of the President, Energy Policy and Planning, April 29, 1977.

[10]"Inexhaustible Energy Resources Planning Study (IERPS)," Vols. 1 and 2, U.S. Dept. of Energy, Review Draft, March 1978.

[11]"Monthly Energy Review," U.S. Dept. of Energy, Energy Information Administration, National Energy Information Center, March 1978.

[12]Aalund, L.R., "U.S. Refining Industry Still Tied to Sweet Crude," *The Oil and Gas Journal*, Vol. 75, No. 42, Oct. 10, 1977, p. 42.

[13]"Energy Alternatives: A Comparative Analysis," Univ. of Oklahoma, Norman, Okla., Science and Public Policy Program, May 1975, pp. 3-31.

[14]Crothers, W.T. and Anderson, C.J., "A Practical Approach to the Introduction of Alternate Automotive Fuels," Lawrence Livermore Lab., Univ. of California, prepared for U.S. Energy Research and Development Administration, Contract W-7405-Eng-48, March 14, 1975.

[15]"Market-Oriented Program Planning Study (MOPPS), Vol. 5: Intermediate Energy Supply Sector," U.S. Dept. of Energy, Assistant Secretary for Energy Technology, Final Rept., Review Draft, Dec. 1977.

[16]"Feasibility of Study of Alternative Fuels for Automotive Transportation," Vol. II, Technical Section, Exxon Research and Engineering Co., prepared for Environmental Protection Agency, June 1974.

[17] "Solar, Geothermal, Electric and Storage Systems Program Summary Document," U.S. Dept. of Energy, Assistant Secretary for Energy Technology, DOE/ET-0041(78), March 1978.

[18] "Charlesworth, G. and Baker, T.M., "Transport Fuels for the Post-Oil Era," Energy Policy, March 1978.

[19] "Should We Have a New Engine?: An Automobile Power Systems Evaluation," Vol. 2, Jet Propulsion Lab., California Inst. of Technology, SP 43-17, Aug. 1975.

[20] Linden, L.A., et al., "Federal Support for the Development of Alternative Automotive Power Systems: The General Issue and the Stirling, Diesel, and Electric Cases," Massachusetts Inst. of Technology, Working Paper, MIT-EL-76-001WP, March 1976.

[21] "A National Plan for Energy Research, Development and Demonstration," Energy Research and Development Administration, ERDA 77-1, June 1977.

[22] "Report from the Alcohol Fuels Task Force," Dept. of Energy, Jan. 18, 1978.

[23] Keller, J.L., Nakuguchi, G.N., and Ware, J.C., "Modification of Methanol Fuel for Highway Vehicles," Wolfsbury Conference, Nov. 1977.

[24] Valencia-Chavez, J.A. and Donnelly, R.G., "Atmospheric Emissions in Energy Source Pollution," AIChE Symposium Series, Vol. 73, p. 312.

[25] Hydrocarbon Process, Dec. 1977, p. 99.

[26] Meisel, S.L., et al., "Gasoline from Methanol in One Step," Chemical Technology, Vol. 6, 1976, pp. 86-89.

[27] Voltz, S.E. and Wise, J.J., "Development Studies on Conversion of Methanol and Related Oxygenates to Gasoline," Energy Research and Development Administration, Final Rept., ERDA Contract E (49-18)-1773, Nov. 1976.

[28] Salzano, F.J. and Braun, C., "Hydrogen Energy Assessment," National Center for Analysis of Energy Systems, Brookhaven National Lab., Upton, N.Y., March 1977.

[29] Escher, W.J.D., "Hyrdogen-Fueled Internal Combustion Engine, A Technical Survey of Contemporary U.S. Projects," Escher Technology Associates, St. Johns, Mich., TEC-75/605, Sept. 1975.

[30] Billings, R.E., "Hydrogen's Potential as a Vehicular Fuel for Transportation," Tenth Intersociety Energy Conversion Engineering Conference, Newark, Del., IEEE Catalog 75 CHO 983-7 TAB, Aug. 18-22, 1975.

[31] Simpson, F.B., Lofthouse, J.H., and Swope, D.R., "Modification Techniques and Performance Characteristics of Hydrogen-Powered IC Engines—State of Art, 1975," Proceedings of the First World Hydrogen Energy Conference, Vol. 3, Univ. of Miami, 1976, pp. 6C/75-6C/95.

[32] "Feasibility Study of Alternate Fuels for Automotive Transportation," Exxon Research and Engineering Co., prepared for Environmental Protection Agency, PB 235 582, June 1974.

[33] Shelef, M., Otto, K., and Otto, N.C., "Poisoning of Automotive Catalysts," Advances in Catalysis (to be published).

[34] "Alcohols—A Technical Assessment of Their Application as Fuels," American Petroleum Inst., Publ. 4261, July 1976; also Hagen, D.L.,

"Methanol as a Fuel," Society of Automotive Engineers Paper, Detroit, Mich., Sept. 26-30, 1977.

[35]Menrad, H., Lee, W., and Bernhardt, W., "Development of a Pure Methanol Fuel Car," Society of Automotive Engineers, Paper 770790, Detroit, Mich., Sept. 26-30, 1977.

[36]Brinkman, N.D., "Effect of Compression Ratio on Exhaust Emissions and Performance of a Methanol-Fueled Single-Cylinder Engine," Society of Automotive Engineers, Paper 770791, Detroit, Mich., Sept. 26-30, 1977.

[37]*International Symposium on Alcohol Fuel Technology—Methanol and Ethanol,* Wolfsburg, Federal Republic of Germany, Nov. 21-23, 1977.

[38]Harrington, J.A. and Pilot, R.M., "Combustion and Emission Characteristics of Methanol," *Society of Automotive Engineers Automotive Engineering Congress,* Paper 750450, Detroit, Mich., Feb. 24-28, 1975.

[39]Hearings, Senate Agricultural Research Subcommittee, Indianapolis, Ind., Dec. 12, 1977.

[40]Hurn, R.W., "Another Look at Alternative Fuel Options," Society of Automotive Engineers, Paper 770759, Milwaukee, Wis., Sept. 12-15, 1977.

[41]Fleming, R.E., "Effect of Fuel Composition on Exhaust Emissions from a Spark-Ignition Engine," U.S. Dept. of Interior, Bureau of Mines, 1970.

[42]Gratch, S., "Advanced Automotive Propulsion—An Overview," *Proceedings, 11th Intersociety Energy Conversion Engineering Conference,* Vol. 1, 1976, p.2.

[43]Boekhause, K.L. and Copeland, L.C., "Performance Characteristics of Stratified Charge Vehicles with Conventional Fuels and Gasoline Blended with Alcohol and Water," *Society of Automotive Engineers International Automotive Engineering Congress,* Special Publ. SP 403, Paper 760197, Detroit, Mich., Feb. 23-27, 1976.

[44]Marshall, H.P. and Walters, D.C., Jr. "An Experimental Investigation of a Coal-Slurry Fueled Diesel Engine," Society of Automotive Engineers, Paper 770795, Detroit, Mich., Sept. 26-30, 1977.

[45]Mitchell, E., Cobb, J.M., and Frost, R.A., "Design and Evaluation of a Stratified Charge Multifuel Military Engine," Society of Automotive Engineers, Paper 680042, Detroit, Mich., Jan. 8-12, 1968.

[46]Hills, F.J. and Schleyerback, C.G., "Diesel Fuel Properties and Engine Performance," *Society of Automotive Engineers International Automotive Engineering Congress,* Paper 770316, Detroit, Mich., Feb. 28-March 4, 1977.

[47]Wagner, T.O., "Economics of Manufacturing Automotive Diesel Fuel," Society of Automotive Engineers, Paper 770758, Milwaukie, Wis., Sept. 12-15, 1977.

[48]Bro, K. and Pedersen, P.S., "Alternative Diesel Engine Fuels: An Experimental Investigation of Methanol, Ethanol, Methane and Ammonia in a D.I. Diesel Engine with Pilot Injection," Society of Automotive Engineers, Paper 770794, Detroit, Mich., Sept. 26-30, 1977.

[49]Barry, E.G., Hills, F.J., Ramella, A., and Smith, R.B., "Diesel Engines for Passenger Cars—Emissions, Performance, Fuel Demand and Refining Implications," *API Refining Department, 42nd Midyear Meeting,* Preprint 06-77, Chicago, Ill., May 10, 1977.

[50]Barry, E.G., Ramella, A., and Smith R.B., "Potential Passenger Car Demand for Diesel Fuel and Refining Implications," *Society of Automotive*

Engineers International Automotive Engineering Congress, Paper 770315, Detroit, Mich., Feb. 28-March 4, 1977.

[51] Pierson, W.R., private communication, Dec. 13, 1977.

[52] "Should We Have a New Engine?: An Automotive Power System Evaluation," Jet Propulsion Lab., California Inst. of Technology, SP 43-17, Aug. 1975.

[53] McLean, A.F. and Davis, D.A., "The Ceramic Gas Turbine—A Candidate Powerplant for the Middle- and Long-Term Future," Society of Automotive Engineers, Paper 760239, Feb. 23-27, 1976.

[54] "Stirling Engine Program," Ford Motor Co., U.S. Energy Research and Development Administration Contractors' Coordination Meeting, Nov. 18, 1975; also "Stirling Engine: Information for News Media," Ford Motor Co., April 14, 1976.

[55] Hart, R., Nasralla, M., and Williams, A., "Formation of Oxides of Nitrogen in the Combustion of Droplets and Sprays of Some Liquid Fuels," *Combustion Science and Technology,* Vol. 11, No. 1-2, 1975, p.57.

[56] LaPoint, C.W. and Schultz, W.L., "Comparison of Emission Indexes within a Turbine Combustor Operated on Diesel Fuel or Methanol," *SAE Transactions,* Paper 730669, Sec. 4, 1973, p. 2449.

[57] Hamilton, W., "Prospects for Electric Cars," General Research Corp., Santa Barbara, Calif., CR-70401, June 1978.

[58] "Fuels and Energy Data: United States by States and Census," Bureau of Mines Information Circular, IC8722, 1976.

[59] "Basic Petroleum Data Book," American Petroleum Inst., Washington, D.C., 1977.

[60] Lamb, J., "The Burning of Boiler Fuels in Marine Diesel Engines," *Institute of Marine Engineers Transactions,* Vol. 60, No. 1, 1948, pp. 1-86.

[61] Smith, J.A., "Operational Experience with Heavy Fuel Oil," Diesel Engineers and Users Association, London, England, Publ. 309, 1966.

[62] Grove, O., "Application of Low-Quality Fuels in Large Bore Diesel Engines Aspects for Future Bunker Fuels and Their Influence on Engine Performance," *Shipcare 1978 Conference.*

[63] "Boiler Fuels in Medium-Speed Engines," *Oil Engine and Gas Turbine,* March 1964, pp. 38-39.

[64] Vogtle, G., "The Use of Heavy Fuel in M.A.N. Trunk-Piston Engines," *The Motorship,* Aug. 1966, pp. 216-220.

[65] Gallois, J., "Moteurs Diesel Rapides A. Haute Suralementation Alimentes Au Fuel Oil Lourd," *10th International CIMAC Conference,* 1973.

[66] "Propulsion: Today's Thrust Is Towards Fuel Economy," *Marine Engineering/Log,* Oct. 1977, pp. 45-53.

[67] Water Removal From High Density Fuel," *The Motorship,* April 1978, pp. 93-97, 112.

[68] Verwoest, K.M. and Colon, F.J., "Residual Fuel Treatment On-Board Ship—Part I," Netherlands Ship Research Center TNO, April 1967.

[69] De Mooy, K.A., Verwoest, K.M., and Van Der Meulen, G.G., "Residual Fuel Treatment On-Board Ship—Part II," Netherlands Ship Research Center TNO, March 1967.

[70] "The Treatment of Heavy Oil," *The Motorship,* Oct. 1965, pp. 318-320.

[71] Coant, P.M., Kohout, F.C., and Lowther, H.V., "Lubrication and Wear Monitoring of High-Performance Marine Diesel Cylinder Oils," *Lubrication Engineering,* Vol. 33, Nov. 1977, pp. 581-589.

[72] Coant, P.M., Davison, A., Fluyt, D., and Kohout, F.C., "Development of Antiwear Cylinder Oil for High Output Crosshead Diesels," *Energy Technology Conference,* American Society of Mechanical Engineers, Paper 77-DGP-10, Houston, Texas, Sept. 18-22, 1977.

[73] "Method and Apparatus for Operating an Internal Combustion Engine with Solid Fuel," U.S. Patent 4077367.

[74] "All About Boilers: Steaming with the Energy Crisis," *Marine Engineering/Log,* Feb. 1974, pp. 35-52.

[75] "Steam: Higher Power, An Expanding Market," *Marine Engineering/Log,* Oct. 1971.

[76] "Raising Steam—Reappraising Steam," *Tanker and Bulker International,* Vol. 4, No. 1-2, Jan./Feb. 1978.

[77] Larsen, G.A., "Propulsion: VAP Turbine Plant: A New Approach to Energy Saving and Low Life-Cycle Costs," *Marine Engineering/Log,* Oct. 1977.

[78] Bonar, J.A., "Fuel Ash Corrosion," *Hydrocarbon Processing,* Aug. 1972, pp. 76-77.

[79] "Upgrading of Coal Liquids for Use as Power Generation Fuels," EPRI Rept. AF-444, Research Project 361-2, Oct. 1977.

[80] McCann, C.R., Demeter, J.J., and Bienstock, D., "Combustion of Pulverized Solvent Refined Coal," The Combustion Inst., Central States Section, April 5-6, 1976.

[81] "Final Report of the General Motors Corporation Powdered Coal-Oil Mixture (COM) Program," U.S. Dept. of Energy, Rept. FE-2267-2, Oct. 1977.

[82] Hodgkins, A.F. and Grams, R.A., "Alternative Fuels for Sea Power," *Proceedings of IMAS,* 1976.

[83] Manning, A.B. and Taylor, R.A.A., "Collidal Fuels," *Institution of Chemical Engineers London Transactions,* 1936.

[84] Brame, J.S.S., "Colloidal or Coal-Oil Fuel," *The Colliery Guardian,* Oct. 28, 1932, p. 822; also Society of Chemical Industry, Chemical Engineering Group, London, Oct. 14, 1932.

[85] Hanse, D.J., "Coal-Oil Mixture as a Diesel Fuel," Masters Thesis, Diesel Dept., North Carolina State College, Raleigh, N.C., 1949.

[86] Marshall, H.P. and Walters, D.C., "An Experimental Investigation of a Coal-Slurry Fueled Diesel Engine," Society of Automotive Engineers, Paper 770795, Sept. 1977.

[87] U.S. Army Fuels and Lubricants Research Lab., San Antonio, Texas (unpublished).

[88] Frainier, L.J. and Colling, P.M., "Ramjet Fuel Analysis Report," Final Rept., Contract Y6E140, Nov. 1977.

[89] Dryer, F.L., "Fundamental Concepts on the Use of Emulsions as Fuels," Aerospace and Mechanical Sciences, Princeton Univ., Princeton, N.J., Rept. 1224, April 1975.

[90] Moses, C.A., "Laboratory Combustor Tests on Fuel Emulsions for the Reduction of Exhaust Smoke from Jet Engines," *Symposium on the Use of Water-in-Fuel Emulsions in Combustion Processes,* U.S. Dept. of Transportation, Transportation Systems Center, Cambridge, Mass., April 1977.

[91] Spadaccini, L.J. and Pelmos, R., "Evaluation of Oil/Water Emulsions for Application in Gas Turbine Engines," *Symposium on Evaporation-Combustion of Fuel Droplets,* American Chemical Society, San Francisco, Calif., 1976.

[92] Valdmanis, E. and Wulfhorst, D.E., "The Effects of Emulsified Fuels and Water Induction on Diesel Combustion," Society of Automotive Engineers, Paper 700736, 1970.

[93] Storment, J. and Coon, C., "Single-Cylinger Diesel Engine Tests with Unstabilized Water-in-Fuel Emulsions," Southwest Research Inst., San Antonio, Texas, Rept. 1206, May 1978.

[94] Moffitt, J.V., "Fuel Economy Evaluation of an Alcohol/Water/Gasoline Mix in a Military SI Engine," U.S. Army Fuels and Lubricants Research Lab., San Antonio, Texas, Rept. 53, Jan. 1975.

[95] Felt, A.E. and Steele, W.A., "Combustion Control in Dual-Fuel Engines," Research and Development Dept., Ethyl Corp.

[96] Karim, G.A. and Klat, S.R., "Experimental and Analytical Studies of Hydrogen as a Fuel in Compression Ignition Engines," American Society of Mechanical Engineers, Paper 75-DGP-19, April 1975.

[97] Hochn, F.W. and Dowdy, M.W., "Feasibility Demonstration of a Road Vehicle Fueled with Hydrogen-Enriched Gasoline," Society of Automotive Engineers, Paper 749105, Aug. 1974.

[98] Parks, F.B., "A Single-Cylinder Engine Study of Hydrogen-Rich Fuels," Society of Automotive Engineers, Paper 760099, Feb. 1976.

[99] Bateman, I. and Wright, M.J., "Further Investigation of Methanol in a Dual Fuel Engine," Riccardo Consulting Engineers Ltd., Sussex, England, Paper SN 19949, July 1975.

[100] Hommer, E., "Methanol as a Substitute Fuel in the Diesel Engine," AB Volvo, Truck Div., Goteborg, Sweden.

[101] Dept. of Mechanical Engineering, Indian Inst. of Technology, Madras, India.

[102] Train, R.E., "Environmental Cancer," U.S. Environmental Protection Agency, Washington, D.C.; also *Science,* Vol. 195, No. 4277, Feb. 4, 1977.

[103] Miller, R.W., "Relationship Between Human Teratogens and Carcinogens," *Journal National Cancer Institute,* Vol. 58, No. 3, 1977.

[104] Pitts, J.N., Jr., Grosjean, D., and Mischke, T.M., "Mutagenic Activity of Airborne Particulate Organic Pollutants," *Toxicology Letters,* No. 65-70, Elsevier Scientific Publishing Co., Amsterdam, Holland, 1977.

[105] Blot, W.J., Brinton, L.A., and Fraumeni, J.F., Jr., "Cancer Mortality in U.S. Countries with Petroleum Industries," *Science,* Vol. 198, No. 51-52-53, 1977.

[106] Henderson, B.E., Gordon, R.J., Mench, H., Soohoo, J., Martin, S.P., and Pike, M.C., "Lung Cancer and Air Pollution in South Central Los Angeles County," *American Journal of Epidemiology,* Vol. 101, No. 6, 1975.

[107] Pihl, R.O., "Lead, Cadmium Linked to Learning Problems," *Science News,* Vol. 112, 1977.

[108] Goldsmith, J.R., "Food Chain and Health Implications of Airborne Lead," State Dept. of Health, Sacramento, Calif., California Air Resources Board Rept. 012, Project 7-083, 1974.

[109] "Air Pollution Health Effects on Children," California Air Resources Board, Staff Rept. 77-20-1, 1977.

APPENDIX A: LIST OF REGISTRANTS

Adams, Martin R.
U.S. Department of Energy

Allsup, Jerry
Oklahoma Department of Energy

Anderson, Carl J.
Lawrence Livermore Laboratory

Bailey, John M.
Caterpillar Tractor Company

Barry, E.G.
Mobil Research and Development Corporation

Baudino, John H.
Atlantic Richfield Company

Bengtsson, Carry
Svensk Metanolutveckling AB

Benjamin, Don
California Department of Transportation

Berger, Jerry E.
Shell Oil Company

Bernhardt, Winfried E.
Volkswagenwerk AG

Bolt, Jay A.
University of Michigan

Borger, J.G.
Pan American World Airways, Inc.

Brateman, Jeffrey H.
Brookhaven National Laboratory

Brewer, Dan
Lockheed-California Company

Bruck, H.W.
University of California at Berkeley

Cart, Eldred N.
Exxon Research and Engineering Company

Cole, Richard B.
Stevens Institute of Technology

Coyne, James
Coyne Chemical Company

Cristofano, Sam M.
City of Santa Clara Public Works

Cruz, Jose Marcio
University of California at Davis

Daley, Ray H.
American Automobile Association

Dean, Gordon W.
U.S. Department of Energy

DePerro, P.L.
Beech Aircraft Corporation

Dickson, Edward
SRI International

DiVacky, Raymond J.
U.S. Postal Service

Dunnam, Blackwell C.
Air Force Aero Propulsion Laboratory

Ecklund, E. Eugene
U.S. Department of Energy

Farmer, Michael
Exxon Research and Engineering Company

Foerster, Willi
Deutsche Shell AG

Fogelberg, Sven-Olof
Svensk Metanolutveckling AB

Fones, Theodore H.
Caterpillar Tractor Company

Francis, Arthur W.
Union Carbide Corporation

Freeman, Jack
Sun Company, Inc.

Fricke, Charles R.
Agricultural Products Industrial Utilization Committee

Furber, Conan P.
Association of American Railroads

Gratch, Serge
Ford Motor Company

Grobman, Jack
NASA Lewis Research Center

Groenewold, G. Michael
Cummins Engine Company

Hamilton, William
General Research Corporation

Haslam, Kent
Lawrence Livermore Laboratory

Heitland, Herbert
Volkswagen DoBrasil S.A.

Hinkle, Jerome
U.S. Department of Energy

Hipkin, Howard
Bechtel National Corporation

Hirao, Osamu
Tokyo University

Hoffman, John G., Jr.
General Electric Company

Hood, Richard
General Motors Technology Center

Houseman, John
Jet Propulsion Laboratory

Hussey, John F.
Solar Turbines International

Husted, Robert A.
U.S. Department of Transportation

Irwin, Richard F.
Chevron Research Company

Jackson, Robert G.
Continental Oil Company

Johnson, Richard T.
University of Missouri at Rolla

Kant, Fred H.
Exxon Research and Engineering Company

Keller, James L.
Union Oil Company

Kemp, Donna R.
Idaho Office of Energy

Koehl, William J.
Mobil Research and Development Corporation

LaFave, Ivan V.
Chicago Bridge and Iron Company

LaRosa, Paul J.
TRW

Leach, H. James
Office of Technology Assessment

Lima, Louis
Centrais Eletricas De Sas Paulo

Longwell, John P.
Massachusetts Institute of Technology

Luchter, Stephen
National Highway Traffic Safety Administration

Maxfield, Daniel P.
U.S. Department of Energy

Meier, Peter
Brookhaven National Laboratory

Melillo, Daniel C.
Ryder Truck Rental, Inc.

Miller, Donald R.
Vulcan Cincinnati, Inc.

Mingle, John G.
Oregon State University

Momenthy, Albert M.
Boeing Commercial Airplane Company

Moses, Clifford
Southwest Research Institute

Myers, Phillip S.
University of Wisconsin

Nakajima, Keitaro
Toyota Technology Center, USA, Inc.

Olivera, Edward S.
Centrais Eletricas De Sas Paulo

Pangborn, Jon B.
Institute of Gas Technology

Panzer, Jerome
Exxon Research and Engineering Company

Parrill, Richard S.
Holly Sugar Corporation

Pasternak, Alan D.
California State Energy Commission

Pefley, Richard
Santa Clara University

Peters, William W.
Ricardo Consulting Engineers

Plassmann, Eberhard
Tuvrheinland

Purohit, G.P.
Jet Propulsion Laboratory

Quillian, R.D.
Southwest Research Institute

Rightmire, Robert A.
SOHIO

Roberts, Justin
Contra Costa Times

Rodebaugh, Dale
San Jose Mercury and News

Sapre, Alex R.
General Motors Technology Center

Satcher, Joel
Holly Sugar Corporation

Sauter, Norman A.
Deere and Company

Singerman, G.M.
Gulf Research and Development Company

Sklar, Scott
New York State Alliance to Save Energy, Inc.

Sosa, Guillermo
Puerto Rico Energy Office

Stone, Charles L.
California Legislature

Surra, Philip
Stanford University

Thompson, Richard C.
J.I. Case Company

Tierney, William T.
Texaco, Inc.

Timbario, Thomas J.
Mueller Associates, Inc.

Timmcke, Wesley E.
Air Products and Chemicals, Inc.

VanLandingham, Earl E.
Office of Aeronautics and Space Technology, NASA

Vann, Leon Guy
California Energy Commision

van Schayk, Christian H.
Motor Vehicle Manufacturers Association

Vieira de Carvalho, Arnaldo
Praia do Flamengo

Waide, Charles H.
Brookhaven National Labratory

Walinchus, Robert J.
Mueller Associates, Inc.

Wallace, R. Eugene
International Harvester Company

Watkins, Lane B.
WED Engineering

White, Herbert
Aerospace Corporation

Wimmer, D.B.
Phillips Petroleum Company

Witcofski, Robert D.
Langley Research Center, NASA

Wolf, A.J.
Teledyne Wisconsin Motor

Young, Thomas C.
Engine Manufacturers Association

Zweig, Robert M.
Pollution Control Research Institute

APPENDIX B:
SELECTED BIBLIOGRAPHY ON ALCOHOL FUEL*

"Alcohol from Babacu," *Amazon News Letter,* Belem, Brazil, March/April 1977. Discusses the use of Babacu trees for alcohol production. Offers novel statistics.

Altsheller, W., et al., "Design of a Two-Bushel per Day Continous Alcohol Unit," *Chemical Engineering Progress,* Vol. 43, No. 9, p. 467. An excellent article on small-scale technology for farmers and alcohol enthusiasts alike. Total fermentation time is eleven hours, which produces 190-proof alcohol. Altsheller, Brown, Stark, and Smith are from John E. Seagrams & Sons, Inc., Kentucky.

American Petroleum Inst., "Alcohols: A Technical Assessment of Their Applications as Fuels," Paper 4261, July 1976. A rather outdated summary of problems with alcohol fuels. Contrasts significantly with the Volkswagen studies. A new report is being worked on.

Anderson, H.H., President, Pacific Rubber Growers, South Pasadena, Calif., Dec. 1977. A letter on the uses of the guayule plant to replace petroleum in the production of rubber.

Anderson, J., "Alcohol Fuels: An Overlooked Answer for Energy?," *The Washington Post,* Sept. 11, 1977. Mr. Anderson outlines the possibilities and political problems.

—"A New Look at Alcohol Fuels," *The Washington Post,* April 9, 1977. The best commentary to date from a member of the national media on the possible uses of alcohol fuels in the U.S. energy economy.

Baratz, B., et al., "Survey of Alcohol Fuel Technology," prepared for the National Science Foundation, Office of Energy Research, The Mitre Corp., McLean, Va., Nov. 1976. A complete survey on government policy on the issue. Should be read.

Barber, S., "The Barber Dual Fuel System," Congressional Information Booklet, Summer 1977. Mr. Barber drove his alcohol-powered car to Washington, D.C. this summer. He awakened the interest of many energy staff on Capitol Hill with his car, which could run on either alcohol or gasoline.

Barr, W.J. and Parker, F.A., "The Introduction of Methanol as a New Fuel into the United States Economy," American Energy Research Co., McLean, Va., March 1976. An excellent review of methanol as an energy source for U.S. and utilities. One of the most thorough economic analyses ever done on methanol fuels.

Beazley, E., "FEA Wants Accounting of 'Forgotten' Methanol Study," *Knoxville Journal,* April 25, 1977. Account of the last study on methanol by the T.V.A.

---

*Prepared by Scott Sklar. See final paragraph, Section VII.G.5, page 212.

Bernhardt, W.E. and Lee, W., "Engine Performance and Exhaust Emission Characteristics of a Methanol Fueled Automobile," *Future Automotive Fuels—Prospects, Performance, Perspective,* Plenum Publishing Co., New York, 1977. A complete technical overview of methanol characteristics. Discusses vapor lock, cold-start, and emissions aspects.

Bernton, H., "Alcohol Fuels: A Major Source of Power Only a Few Years from Here," *Environmental Action Bulletin,* Oct. 29, 1977. An excellent, accurate summary by a reporter who has been investigating this issue for some time.

"Biomass Energy for Hawaii," Stanford Univ./Univ. of Hawaii Biomass Energy Study Team, Feb. 1977. A four-volume study concerning marine plantations, energy farms, sugar operations, municipal wastes, etc. Contains graphs, statistics, diagrams and, in addition, is well written. One of the best biomass studies.

All annotations will be omitted from here on and may be obtained from the author in the larger annotated bibliography.

Black, W.E., "Alternative Organizational and Marketing Arrangements for Marketing Biomass," *Conference on the Production of Biomass from Grains, Crop Residues, Forages, and Grasses for Conversion to Fuels and Chemicals,* Kansas City, Mo., March 2, 1977.

Blaser, R.F., "Heat Balanced Cycle," U.S. Naval Academy, Div. of Engineering and Weapons, Annapolis, Md., 1974.

Boatwright, D., "Synthetic Fuels Car Development," submitted to the California Assembly Rules Committee, Aug. 1, 1977.

—"Plan for Action for Phase II—Biomass Farm Feasibility Assessment," Sept. 1977.

Borgenstam, C.,"Racing Fuels," *Bugantics (England),* Vol. 40, No. 3, p. 26.

Brackett, A.T., et al., "Indiana Grain Fermentation Alcohol Plant," State House, Indianapolis, Ind., No. 336, 1976.

Brinkman, N.D., "Vehicle Evaluation of Neat Methanol—Exhaust Emissions, Fuel Economy, and Driveability," General Motors, GMR 2402, April 18, 1977.

Bryce, A.J., "A Research and Development Program to Assess the Technical and Economic Feasibility of Methane Production from Giant Brown Kelp," American Gas Association, Chicago, Ill., 1977.

Burke, D.P., "CW Report," *Chemical Week,* Sept. 24, 1975, p. 33.

Calvin, M., "Green Factories," *Chemical and Engineering News,* March 20, 1978, p. 30.

Carbide to Expand Taft 1A Facility, Lift Ethanol Prices," *The Wall Street Journal,* May 27, 1977.

"Catalyst Key to Cheaper Coal Liquification," *The Oil Daily,* Jan. 17, 1978, p. 3.

Chang, C.D. and Silvestri, A.J., "The Conversion of Methanol and Other O-Compounds to Hydrocarbons over Zeolite Catalysts," *Journal of Catalysts,* Vol. 57, No. 2, May 1977.

"Cleveland Discol," advertisement, *Motor Sport,* Aug. 1963, p. 578.

"Coal and Climate: A Yellow Light on CO_2," *Science News,* July 30, 1977.

Cohn, R.D., Kant, F.H., et al., "Feasibility Study of Alternative Fuels for Automobile Transportation," NTIS, Springfield, Va., 1974.

Colucci, J.M. (ed.), "Summary of GM Research: Outlook on Methanol," General Motors Technical Center, Warren, Mich., March 10, 1976.

Committee on Renewable Resources for Industrial Materials (CRRIM), National Academy of Sciences, Washington, D.C., 1976.

"Consultant: Gasohol Popular Topic," *Omaha World-Herald,* Jan. 25, 1978.

Cowper, W., "The Alcohol Fuels," *Energy and Earth Machine Magazine,* W.W. Norton & Co., 1976.

Cray, C.L., *Gasohol Seminar,* Rio de Janeiro, Brazil, Midwest Solvents, Kansas, 1976.

Curry, R.F., "Alcohol Fuels," *The American Motorist,* Sept.-Oct. 1977.

—"Future May Be Now for Alcohol Fuels," *American Motorist,* May 1977.

Curtis, C., "Commission on Increased Industrial Use of Agricultural Products," Senate Doc. 45, U.S. General Post Office Rept. PL-540, Washington, D.C., 1957.

Doring, C., "Can White Lightening Solve the Energy Crises?," Bethesda, Md., Nov. 1976.

Dover, H., "Joint Resolution #42: Direct Relation of the Feasibility of Conversion of Wheat and Barley into Methanol," *Montana WIFE,* Oct. 1977.

Duckworth, E., "Gasohol Controversy Still Brewing—Has Backers and Blockers," No. 11, Oct. 1977.

Ecklund, E.E., "Gasohol—An Alcoholic Dilemma?," *Nebraska Midwest Regional Alcohol Conference,* Nov. 1, 1977.

—"Avoiding Energy Catastrophe with Evolutionary Fuels Derived via Systems Technology," Energy Research and Development Administration, Aug. 31, 1977.

Ell, F.J., "Garbage Gas: Fuelish Idea?," *The Billings Gazette,* Oct. 29, 1977.

Ellis, T.J., "Should Wood Be a Source of Commercial Power?," *Forest Products Journal,* Oct. 1975.

Ellis, W.N., "A.T.: The Quiet Revolution," *Bulletin of Atomic Scientists,* Nov. 1977, p. 25.

"Energy Department Backs Smaller Ways of Conserving," *The Washington Post,* Feb. 19, 1978.

Enzymatic Hydrolosis of Cellulosic Wastes to Fermentable Sugars and the Production of Alcohol," Pollution Abatement Div., U.S. Army Natick Research and Development Command, Jan. 29, 1976.

ERDA Dept. Memo., "Position Paper on the Status of Using Alcohol/Gasoline Blends in Highway Vehicles," Energy Research and Development Administration, July 1977.

—"Economic Feasibility of Fuel Grade Methanol from Coal," Office of Commercialization, Energy Research and Development Administration, Wilmington, Del., 1976.

Estes, E., remarks at Detroit Athletic Club, General Motors Corp., May 6, 1977.

Faltermayer, E., "The Clean Synthetic Fuel That's Already Here," *Fortune,* Sept. 1975.

Faulkner, D., "Staff Report on Methanol Fuels," State of California Air Resources Board, Sept. 21, 1975.

Federal Register, Commodity Credit Corp., "Consideration of Industrial Hydrocarbons Pilot Program," Vol. 42, No. 203, Oct. 20, 1977.

Finn, R.K. and Ramalingham, A., "The Vacufern Process: A New Approach to the Fermentation of Alcohol," *Biotechnology and Bioengineering,* Vol. 19, No. 4, 1977.

Fleming, A., "Realistic Outlook for Liquid Fuel Resources," *The Washington Star,* Nov. 4, 1977.

Flint, J., "Fuel for Debate: Gasoline or Alcohol," *The New York Times,* Jan. 29, 1978.

"Foe Questions Need, Gasohol Plant Profit," *Sunday Journal and Star,* Lincoln, Neb., May 29, 1977.

Fourth International Symposium on Automotive Propulsion Systems, Vol. II, Sessions 4 and 6, Energy Research and Development Administration, Washington, D.C., 1976.

Fricke, C.R., letter, *Nebraska Midwest Regional Conference,* Jan. 3, 1978.

"Fuel of the Future?," *Automotive Engineering,* Dec. 1977, p. 48.

Gandelman, J., "Spain, Gas Sight for Gas Substitute," *Christian Science Monitor,* Aug. 15, 1977.

—"Road-Testing New 'Gas' High in Spain," *Christian Science Monitor,* Aug. 22, 1977.

"Gasohol—Another Answer?," *Motour,* Oct. 1977, p.11.

"Gasohol as Future Fuel," *San Francisco Chronicle,* Dec. 11, 1977.

"Gasohol Backer Sees Economic Benefits," *Sunday Journal and Star,* Lincoln, Neb., May 29, 1977.

"Gasohol—Current Status and Potential for the Future," Illinois Farm Bureau and Commodities Div., Research Rept., Feb. 1978.

"Gasohol Gaining in D.O.E.," *The Energy Daily,* Washington, D.C., Feb. 7, 1978, p. 1.

"Gasohol: Let's Get It Straight," editorial, *The Denver Post,* Feb. 15, 1978, p. 16.

"Gasohol Sold Like Hotcakes at Illinois Pump," *San Francisco Sunday Examiner,* Dec. 11, 1977, p. 16B.

Goldstein, I.S., "The Potential for Converting Wood into Plastics and Polymers or into Chemicals for the Production of These Materials," North Carolina State School of Forest Resources, Dept. of Wood and Paper Science, Raleigh, N.C., NSI-Rann Rept., 1974.

Graham, R.W., "Fuels from Crops: Renewable and Clean," *Mechanical Engineering,* May 1975.

Green, F.L., "Energy Potential from Agricultural Field Residues," General Motors Technical Center, Warren, Mich., June 1975.

Hagen, D.L., "Methanol as a Fuel: A Review with Bibliography," Society of Automotive Engineers, Paper 770292, 1977.

—"Methanol: Its Synthesis, Uses as a Fuel, Economics and Hazards," NTIS, Commerce, Va.

Hagey, G., et al., "Methanol and Ethanol Fuels—Environmental, Health and Safety Issues," Dept. of Energy and Mueller Associates, Baltimore, Md., 1977.

Hammond, A.H., "Methanol at M.I.T.: Industry Influence Charged in Project Cancellation," Science, Nov. 1975.

—"Photosynthetic Solar Energy: Rediscovering Biomass Fuels," Science, Dec. 1976.

—"Alcohol: The Brazilian Answer to the Energy Crises," Science, Dec. 1976.

Hammond, D.C., Jr., et al., "Combustion of Methanol in an Automotive Gas Turbine," Resource Recovery and Energy Review, Wakeman-Walworth Inc., Summer 1977.

Harney, B.M., "Methanol from Coal—A Step Toward Energy Self Sufficiency," Energy Sources, Vol. 2, No. 3, 1975.

Hayes, D., "The Coming Energy Transition," The Futurist, Oct. 1977.

—"We Can Use Solar Now," The Washington Post, Feb. 26, 1978.

Healey, J., "National Organization Created to Promote Use of Gasohol," Des Moines Register, Jan. 25, 1978.

Hieronymus, W., "Brazil Tries Mixing Alcohol from Sugar with Gasoline to Reduce Its Oil Imports," The Wall Street Journal, Nov. 28, 1977, p. 30.

Hilden, D.L. and Parks, F.B., "A Single Cylinder Engine Study of Methanol Fuel—Emphasis on Organic Emissions," Society of Automotive Engineers, Feb. 23, 1976.

Hildenbrand, B., "Methanol: An Automobile Fuel in Our Future?," General Motors Research Lab., Jan.-Feb. 1977.

Humphreys, G.C., et al., "The I.C.I. Methanol Process—Past, Present and Future," Chemical Economy and Engineering Review, Nov. 1974, p. 26.

"Industrial Alcohol," War Changes in Industry Series, U.S. Tariff Commission, Jan. 1944.

"Industrial Alcohol from Wood and Agricultural Residues," Forpridecom Technical, Laguna, Phillipines, No. 176, April 1977.

Ingamells, J.C. and Lindquist, R.H., "Methanol as Motor Fuels," Chevron Engine Fuels Div., 1972.

Inman, R.E., "Summary: Silvacultural Biomass Farms," Mitre Corp., Metrek Div., May 1977.

"Integrated Solar Assisted Food and Ethanol Food Program," Domestic Technology Inst., Lakewood, Colo., Jan. 1978.

Intertech Energy Co., "Pictorial Flow Sheet for Algahol Process," Bethesda, Md., 1977.

Jacobs, P.B. and Newton, H.P., "Motor Fuels from Farm Products," USDA, Washington, D.C., Miscellaneous Publ. 327, 1938.

Jamison, R.L., "Trees as a Renewable Energy Source," Weyerhaeuser Co., Energy Management Group, Tacoma, Wash., Jan. 1977.

Janeway, E., letter to R. Janeway on alcohol fuels, April 13, 1977.

—"The 10% Solution (to the Farm Strike)," San Francisco Sunday Examiner and Chronicle, Jan. 15, 1978.

Jarvis, P.N., "Methanol as a Gas Turbine Fuel," Engineering Foundation Conference, July 8, 1974.

Johnson, R.T., "Methanol Gasoline Blends—Future Automotive Fuels," Dept. of Mechanical Engineering, Univ. of Missouri, Rolla, *Energy Spring*, 1977, p. 27.

Jonchere, J.P., "Methanol Seen as Hydrogen Source," *Oil and Gas Journal*, Petroleum Publishing Co., June 14, 1976.

Keim, C.R., "Sweeteners from Starch," *Sugar y Azucar*, Feb. 1978, p. 53.

Keller, J.L., et al., "Modification of Methanol Fuel for Highway Vehicles," Union Oil Co. of Calif., Blea, Calif., 1977.

Keller, L.J., "Statement Prepared for the Committee on Appropriations," Feb. 2, 1978.

Kemp, C.C. and Szego, G.C., "Energy Forests and Fuel Plantations," Chemtech, May 1973.

Kinsey, R.D., "Garbage Power," *Catalyst*, Vol. V, Nov. 2, 1976, p. 23.

Kirik, M., "Alcohol as an Alternate Fuel," *The Ontario Digest and Engineering Digest*, Sept. 1977.

Klass, D.L., "Biomass and Waste Production as Energy Resources," *Update Energy*, Fall 1977.

Kostick, J., et al., "The Use of Absorbed Cellulose in the Continuous Conversion of Cellulose to Glucose, " *Journal of Polymer Science*, 1971.

Kramon, G. and Sanders, M. (eds.), "An Integrated Plan for the Conversion of Solid Waste to Energy in Santa Clara County," Inst. for Energy Studies, Santa Clara Univ., Santa Clara, Calif., Nov. 1975.

Lave, L.B. and Freeburg, T., "Health Effects of Electricity Generation from Coal, Oil, and Nuclear Fuel," *Nuclear Safety*, Vol. 14, No. 5, Sept.-Oct. 1973, p. 409.

Lawrence Livermore Lab. (transl.), "On the Trail of New Fuels," Federal Ministry for Research and Technology, West Germany, UCLA, Livermore, Calif., Aug. 29, 1975, 585 pp.

Lincoln, J.W., *Methanol and Other Ways Around the Gas Pump*, Gateway Publishing, Charlotte, Vt., 1977.

Lindsey, D.S., "Grain Alcohol Study Prepared for the Indiana Department of Commerce," Long Rock J.V. Study Group, July 2, 1975.

Lindsley, E.F., "Alcohol Power: Can It Help You Meet the Soaring Cost of Gasoline," *Popular Science*, Vol. 206, April 1975.

Lipinsky, E.S., "Sugar Cane vs. Corn vs. Ethylene as Sources of Ethanol for Motor Fuels and Chemicals," Batelle Columbus Labs., Columbus, Ohio, June 23, 1977.

—"The Prospects for Fuels from Biomass," *I.E.C.E. Conference*, Washington, D.C., Aug. 30, 1977.

—"Systems Study of Sugar Cane, Sweet Sörghum, Sugar Beets, and Corn for Fuels and Chemicals," Energy Research and Development Administration, Contract W-7405ENG92, March 15, 1977.

Longwell, J.P. and Most, W.J., "Single-Cylinder Engine Evaluation of Methanol—Improved Energy Economy and Reduced NO$_x$," Society of Automotive Engineers, New York, 1975.

Lubell, D., Environmental Technology, Inc., Bethpage, N.Y., Nov. 1977.

Ludvigsen, K., "Alcohol Comes Back to Power Your Car," *Mechanics Illustrated*, Jan. 1978.

Lugar, D., "Economic Feasability of Gasohol," hearing before the Senate Committee on Agriculture, Nutrition, and Forestry, Subcommittee on Agricultural Research and General Legislation, GPO 22-763, 1978.

Mandels, M., et al., "Disposal of Cellulosic Waste Materials by Enzymatic Hydrolysis," *Army Science Conference Proceedings,* Vol. 3, June 1972.

Mariani, C., *Appunti di Storia,* Italy.

Marvin, M. (ed.), "Methanol Output Seen Topping 100,000 Tons/Year by 2000," *The Oil Daily*, Feb. 24, 1978.

McCloskey, J.P., "Methanol as a Replacement for Gasoline," *Energy Sources,* Vol. 2, No. 1, 1975.

McElheny, V.K., "Multi-Fuel Car Engine Gets New Tests," *The New York Times,* Sept. 12, 1977.

Meisel, S.L., et al., "Gasoline from Methanol in One Step," *Chemical Technology,* Feb. 1976.

—"Recent Advances in the Production of Fuels and Chemicals over Zeolite Catalysts," *Leo Friend Symposium,* Chicago, Ill., Aug. 30, 1977.

Menrad, H., Lee, W., and Bernhardt, W., "Development of a Pure Methanol Car," *Society of Automotive Engineers Passenger Car Meeting,* Paper 770790, Detroit, Mich., Sept. 1977.

Methyl Fuel, Clean Energy, Wentworth Bros. Inc., Cincinnati, Ohio, 1976.

Miller, D., "Fermentation of Ethyl Alcohol," *Biotechnology and Bioengineering Symposium, USDA Proceedings of the 7th National Conference on Wheat Utilization Research,* Paper 6307-312, Manhattan, Kansas, Nov. 1971.

"Mobil Says Process Bolsters Coal-to-Gasoline Potential," *Coal Week,* Jan. 23, 1978, p. 10.

Mohr, B.J. and Scheller, W.A., "Net Energy Analysis of Ethanol Production," American Chemical Society, Preprints 29 and 21, Vol. 2, 1976.

"Motor Alcohol," *Bulletin of the U.S. Food Administration,* Honolulu, Hawaii, July 30, 1918.

Murnane, T., "California Waste-Wood Co-Generation Planned," *Energy User News,* Vol. 3, No. 4, Jan. 23, 1978, p. 1.

Myerson, E.J., Dept. of Energy Memo. to T.E. Noel, Nov. 4, 1977.

"Oil: Autos with Less Pollution Seen," *The Oil Daily,* Nov. 29, 1978, p. 8.

Omang, J., "How to Mix Alcohol Automobiles or, There's Some Vodka in Your Gas," *The Washington Post,* Oct. 20, 1977.

—"Mr. Bell Had a Solution to the Energy Crises," *The Washington Post,* Dec. 15, 1977, p. a54.

—"Study Sends Cool Breeze Toward Solar Energy Backers," *The Washington Post,* Dec. 18, 1977.

Pain, B. and Phillips, R., "The Energy to Grow Maize," *New Scientist,* May 1975.

Peterson, "Mobil's Methanol to Gasoline Process," news release, Paulsboro, N.J., Jan. 1978.

"Planting Oil," Banco Real, Brazil, April 11, 1977.

Pleet, S.W., *Alcohol: A Fuel for Internal Combustion Engines,* Chapman and Hall, London, 1949.

"Political Economy of Fuel Alcohol," Hudson Inst., Croton-on-Hudson, N.Y., Jan. 10, 1978.

Posner, H.S., "Biohazards of Methanol in Proposed New Uses," *Journal of Toxicology and Environmental Health,* 1975, p. 153.

Pouring, A.A., Blaser, R.F., Keating, E.L., and Renkin, B.H., "Influence of Combustion with Pressure Exchange on the Performance of Heat Balanced Internal Combustion Engines," Society of Automotive Engineers, Feb. 28, 1977.

Povich, M.J., "Some Limitations of Fuel Farming," General Electric Co., Schenectedy, N.Y., 16 pp.

"President's Advisory Panel on Timber and the Environment (PAPTE)," General Post Office, Washington, D.C., Rept. 1973.

Proceedings, Forest and Field Fuels Symposium, Biomass Energy Inst., Canada, Oct. 1977, 350 pp.

"Production of Ethyl Alcohol and Nutrients from Municipal Water and Solid Wastes, Agricultural Wastes and Agricultural Products," Solar Fuel Co., Mankato, Minn., March 1978.

"Prospects for Fuels from Biomass," *IECE Conference,* Washington, D.C., Aug. 30, 1977.

Pumental, et al., "Food Production and the Energy Crises, *Science,* Nov. 1973.

Raloff, J., "The Third World Needs Energy Too," *Science News,* Oct. 8, 1977.

Raphael Katzen Assoc., "Chemicals from Wood Waste," USDA, Forest Service, Forest Products Lab., Madison, Wis., Dec. 24, 1975.

Reed, T., "Alcohol Fuels—The Clean Renewable Substitute for Petroleum," *International Symposium on Energy Sources and Development,* Barcelona, Spain, Oct. 19, 1977.

—"Biomass Energy Refineries for Production of Fuel and Fertilizer," *Applied Polymer Symposium,* No. 28, 1975.

—"Comparison of Methanol and Methanol Blends," Massachusetts Inst. of Technology, Energy Lab., Cambridge, Mass.

—"Efficiencies of Methanol Production from Gas, Coal, Waste or Wood," *ACS Net Energetics of Integrated Synfuel Systems Symposium,* April 4, 1976.

—"Improved Performance of Internal Combustion Engines Using 5-30% Methanol in Gasoline," *Inter-Society Energy Conversion Engineering Conference,* Paper 749104, Aug. 1974, p. 952.

—"The Potential Impact of Widespread Use of Wood and Alcohols," *Symposium on Alcohol as Alternative Fuels,* Ontario, Canada, Nov. 19, 1976.

—"The Use of Alcohols and Other Synthetic Fuel in Europe from 1930 to 1950," *AIChE Meeting,* Boston, Mass., Sept. 1975.

—"When Oil Runs Out—A Survey of Our Primary Energy Sources and the Fuels We Can Make from Them," *Capturing the Sun Through Bioconversion Conference,* Washington, D.C., and *Swedish Methanol Conference,* Swedish Royal Academy of England.

Reed, T. and Lerner, R.N., "Methanol: A Versatile Fuel for Immediate Use," *Science,* 1973, p. 182.

Republican Conference: Alcohol: The Renewable Fuel from Our Nation's Resources, U.S. Senate, Oct. 1977.

Roberts, J., "Experimental Car," *Contra Costa Times,* booklet, 1977.

Roemmec Guarantee Fuels, Inc., "Energy Pellet Booklet," Independence, Kansas, 1977.

Rohan, B. and Ross, N., "One Answer to Arab Oil Control," *The Free Press,* March 20, 1977.

Roosevelt, F.D., "Report of the Energy Resources Committee to the Natural Resources Committee," Washington, D.C., Jan. 1939.

Rosenbluth, R.F. and Wilke, C.R., "Comprehensive Studies of Solid Wastes Management: Enzymatic Hydrolysis of Cellulose," Sanitary Engineering Research Lab., Berkeley, Calif., Dec. 1970, p. 79.

Saeman, J.F., "Energy and Materials from the Forest Biomass," U.S. Forest Products Lab., Madison, Wis., Jan. 1977.

Scheller, W.A., "The Development of a High Protein Isolate from Selected Distillers By-Product," Univ. of Nebraska, Final Rept. on NSF Grant, AER 74-10456 A01, July 1975.

—"Gasohol: Fuel and Food for the Future," Univ. of Nebraska, Lincoln, Neb., 1977.

—"Use of Ethanol Gasoline Mixtures for Automobile Fuel," Dept. of Chemical Engineering, Univ. of Nebraska, Jan. 25, 1977.

Scheller, W.A. and Mohr, B.J., "Gasoline Does Too Mix with Alcohol, " Chemtech, Oct. 1977.

—"Net Energy Analysis of Ethanol Production," American Chemical Society, Fuel Chemicals Div., 1976.

Schinto, J., "Alcohol for Gasoline," *The Progressive,* Nov. 1977, p. 46.

Schons, J.J., "Change Gas to Methanol, Trim Costs," *Oil and Gas Journal,* Petroleum Publishing Co., Oct. 11, 1976.

Schroeder, D., "Instead of Wasting Nonrenewable Energy, Why Don't We Exploit Our Wastes?," *Science Forum,* June 1976, pp. 3-6.

"Seameth: Study Summary—A Marine Methanol Plant," Continental Oil Co. (CONOCO), 1977.

"Seven-Year Test Program Okays Alcohol as Motor Fuel," *Western Pennsylvania Motorist,* A.A.A., Feb. 1978.

Shepard, T., "Methanol Energy Key, Expert Says," *The Champaign-Urbana News Gazette,* Aug. 14, 1977.

Sklar, S., "The Chevron Sheet in Perspective," prepared for U.S. Senate Staff, Nov. 1977.

—"Evaluation of Wilburn Article in the January 31, 1978 *The Oil Daily,*" Feb. 10, 1978.

Soloman, B., "Moonshine and Motor Cars: Alcohol Fuels Come of Age," *The Energy Daily,* Washington, D.C., Vol. 5, No. 205, Oct. 28, 1977.

"State of California: Energy from Waste," Joint Committee on Job Development, State Legislature, Dec. 10, 1976.

"State of Maine: Maine Methanol: Collected Working Papers on the Production of Synthetic Fuel from Wood," Office of Energy Resources, Augusta, Maine, March 21, 1975.

"Statement of Americans for Energy Independence," *Senate Appropriations Committee Hearings on Alcohol Fuels,* Jan. 31, 1978.

Stiles, A.B., "Methanol: Past, Present, and Speculation on the Future," *Journal Review,* May 1977.

Stone, C.L., testimony before the Senate Appropriations Committee, Jan. 31, 1978.

"Support the Gasohol Study," *The Sunday Denver Post,* Jan. 8, 1978.

"Survey of Bioconversion Processes for Biomass," *2nd Pacific Chemical Engineering Conference (PACHEC),* Denver, Colo., Aug. 1977.

ter Horst, J.F., "Gas and Alcohol Pose Another Problem," Universal Press Syndicate, *The Detroit News,* Editorial, 1977.

—"Gasohol—No Fuel Like an Old Fuel," *Los Angeles Times,* Nov. 8, 1977.

Thomas, C.O., et al., "Methanol from Coal, Fuel and Other Applications," Oak Ridge Associated Univ., Inst. for Energy Analysis, Feb. 1976.

Tompkins, A., "Energy from Marine Biomass Project," American Gas Assoc., Program Review, Feb. 1, 1978.

Toyota Motor Co., "Effects of Alcohol Reformed Gas Supplement to an Automotive Gasoline Engine," Japan, Sept. 1977.

Tsao, G.T., "Utilization of Grain and Crop Residues for the Production of Fuel and Chemicals," Purdue Univ., Jan. 31, 1978.

"U.S. Department of Energy Alcohol Fuels Program Plan," Task Force on Alcohol Fuels, Internal Doc., Jan. 31, 1978.

U.S. Energy Research and Development Administration, contracts for *ERDA Highway Vehicle Systems Contractors Coordination Meeting,* Dearborn, Mich., Oct. 6, 1977. Papers include: Bartlettsville, Oklahoma Energy Research Center, Univ. of Michigan, Univ. of Santa Clara, Southwest Research Inst. and Standard Oil Co., Union Oil Co. of Calif., U.S. Army Fuels and Lubricants Lab.

U.S. Forest Service, "The Feasibility of Utilizing Forest Residues for Energy and Chemicals," March 1976.

U.S. Senate Dear Colleague Letter to Secretaries Schlesinger and Bergland, signed by 28 U.S. Senators, Oct. 12, 1977.

U.S. Tariff Commission, "Industrial Alcohol," *War Changes in Industry Series,* Jan. 1944.

Vetter, R.L., "Cornstalk Grazing and Harvested Crop Residues for Beef Cows," Iowa State Univ. Cooperative Extension Service, A.S. Leaflet R186, July 1973.

Vogelbusch. Bohler Bros., booklet, P.R. data sheets, Texas, 1978.

Volkswagen:

Berhhardt, W., "45 VW and Audi Vehicles Test Methanol-Petrol Fuel Mixture," Washington, D.C., April 17, 1977.

Volkswagen International Symposium on Alcohol Fuel Technology, Vols. I-III, Wolfsburg, Germany, Nov. 1977.

"Volkswagen: On the Trail of New Fuels," *Alternative Fuels for Motor Vehicles,* 1977, 20 pp.

Wigg, E.E., "Methanol as a Gasoline Extender: A Critique," *Science,* Vol. 186, No. 4166, Nov. 29, 1974.

Wilburn, G., "Gasohol Likely to Produce More Problems Than Benefits for Petroleum Marketers and Consumers," *The Oil Daily,* Jan. 31, 1978.

Wilkie, H.F. and Kolachov, P.J., "Food for Thought," Indiana Farm Bureau, Indianapolis, Ind., 1942.

Wittwer, S.H., "Maximum Production Capacity of Food Crops," Michigan Agricultural Experimental Station, Journal Article 6622.

Wolfe, E., et al., "Emissions and Fuel Economy of a Stratified Charge Engine Operating on a Gasoline/Methanol Blend," National Science Foundation, NTIS, 1975.

Wyss, A., "Major New Fuel Seen for Industries," *Journal of Commerce,* Oct. 17, 1977.

Yuen, P.C., "Molasses-Derived Ethanol + Gasoline Blend," Univ. of Hawaii Natural Energy Inst., User Acceptenace Program, Nov. 1977.